CELEBRATING THE CULTURAL LANDSCAPE HERITAGE OF MILLS COLLEGE

USING THE PAST TO INFORM CAMPUS PLANNING AND DESIGN

Vonn Marie May

Robert Sabbatini

with

Karen Fiene

CENTER FOR THE BOOK

Mills College

Front Cover

LAKE ALISO, CIRCA 1916
Lantern procession and bonfire.

Back Cover

TEMPLE TO MUSIC
Proposed design scheme for the pond near the Music Building.

ISBN: 978-0-9648938-6-3

First published in January 2008

Publisher

Center for the Book
Mills College
5000 MacArthur Boulevard
Oakland, CA 94613-1301

www.mills.edu/academics/library/ctr_book

Printed by the University of California Printing Services

The paper used in this book is made of 100% recycled materials, 50% post-consumer waste, processed chlorine free, ancient forest friendly.

This book is printed with soy-based ink.

TABLE OF CONTENTS

Celebrating the Cultural Landscape Heritage of Mills College offers an essential framework that honors the College's historic legacy and will help guide long-range stewardship of the campus. Documenting the expansion and development of the campus from its mid-19th-century origins to the present, this plan examines Mills' cultural landscape and nationally prominent architecture in the context of an overriding commitment to women's education. This in-depth study of the campus' origins, landscapes, and designers provides a contextual chronology reflecting the College's distinctive beauty and character.

Mills' verdant 135-acre campus—the country's largest liberal arts college campus within an urban setting—is an exceptional environment for world-class academic programs. The stately row of London plane trees lining Richards Road welcomes all who enter campus to discover the oasis of spacious meadows, lush foliage, and historic architecture. The College honors this extraordinary setting by implementing sustainable practices in restoring the natural environment and in constructing and renovating buildings. From the buildings designed by Julia Morgan in the early 20th century, to the Natural Sciences Building completed in 2007, Mills provides a model for 21st-century campus innovation, recently evidenced in new facilities that meet rigorous standards for Leadership in Energy and Environmental Design (LEED) "green buildings."

Among the architects, landscape architects, botanists, and planners who helped to shape the campus are Walter Ratcliff, Jr., campus architect (1920-1940), who worked closely with Howard McMinn, professor of botany, and Howard Gilkey, landscape architect. Mills still benefits from its place as a leader during the Golden Age of Horticulture (1918-1950) when the campus served as a botanical testing ground with 2,500 native trees from 41 species.

We are deeply grateful to the Getty Foundation for selecting Mills to receive a 2006 Campus Heritage grant in generous support of this vital Landscape Heritage Plan, which greatly enriches our understanding of the many influences that have shaped Mills' legacy. As we advance in thoughtful planning and growth for future generations we hope you will be inspired to join and support the Mills community as an invaluable treasure to the City of Oakland, the San Francisco Bay Area, and California.

Janet L. Holmgren

Janet L. Holmgren
President, Mills College
January 2008

OVERVIEW

The following provides an overview of the contents of each book section.

1.
SETTING THE STAGE

Presents the purpose of the study and a summary of the document.

2.
EXPLORING THE HISTORY OF THE CAMPUS

Discusses two periods of historic significance in campus development that contributed the most character-defining features visible on the campus today: the first from 1868 to 1916 (the English country estate and the picturesque era) and the second from 1916 to 1949 (the modern campus).

3.
APPLYING THE PAST TO THE PRESENT

Describes the cultural landscape assessment process and its application in two areas of the campus: the first, the Oval and area surrounding El Campanil, and the second, the internal campus gateway formed by Richards Road and Kapiolani Road.

4.
APPRECIATING THE LEGACY

Summarizes the Mills College course curriculum developed for the project:

- *Mills College: Architecture and Heritage*, taught by Professor of History Bertram Gordon
- *The Greening of Mills College: How to Conserve Resources for a Sustainable Future*, taught by Assistant Professor of Biology, Susan Spiller

and four public lectures:

- "The Design of Mills College—A Campus Rooted in Hope," presented by Helen Lefkowitz Horowitz, Sydenham Clark Parsons Professor of History, Smith College
- "Mills Horticultural Glory Years: 1915–1940," presented by Phoebe Cutler, landscape historian
- "Reading the Cultural Landscape of Mills College: Historic Findings from the Getty Foundation Campus Heritage Project," presented by Vonn Marie May, cultural landscape specialist
- "Mills College: The Architectural Legacy of Walter H. Ratcliff, Jr.," presented by Woodruff Minor, architectural historian

To be transported from the limits of the circumscribed town to God's great, beautiful out-of-doors, to the freedom and joy of the hills.

—CYRUS T. MILLS, CIRCA 1870

0-1

MILLS HALL AND THE SAN FRANCISCO BAY, 1870s

Early view of the campus looking southwest over Mills Hall to the San Francisco Bay beyond.

1.
SETTING THE STAGE
HOW THIS STUDY WILL INFORM AND GUIDE THE CAMPUS' FUTURE

When Cyrus and Susan Mills situated Mills College in the Oakland hills, they chose a landscape characterized by natural California elements. Streams lined with oaks and riparian vegetation traversed the property, winding from the hills above the campus down to San Francisco Bay. In 1871, Cyrus and Susan completed Mills Hall, a commanding piece of architecture on the region's most prominent hill. This inspired setting created not just a scenic tableau, but an icon for students, faculty, and visitors. From its beginnings, the Mills campus has embodied the American frontier spirit: independent, experimental, and striking a harmonious balance between humanity and nature. The founders of Mills College were dedicated to the moral imperatives of education for young women and a celebration of the natural world.

Over the years, Cyrus and Susan Mills and those who followed shaped this setting with plantings of exotic and native plants as well as with distinguished architecture. They worked in partnership with notable landscape architects, botanists, and architects whose influence extended beyond the campus.

Celebrating the Cultural Landscape Heritage of Mills College explores how these great innovators fashioned the campus environment in a way that built on its exceptional natural resources. Thus, the Mills College campus consists of a distinct layering of built landscape design upon the natural topography, reflecting periods of institutional and national history through the styles and theories that informed them. The cultural landscape reveals the philosophy of the founders and their mission to foster innovative education for women.

Celebrating the Cultural Landscape Heritage of Mills College documents the history of the Mills College campus through two periods of historic significance, identifying the major natural and built characteristics that define the Mills College campus. This study provides a framework to guide future development in balance with historic preservation, furthering the College community's appreciation of the campus and its cultural landscape values.

But some fondness kept drawing them back to the district above the San Leandro road, where little capes of foothills reach down upon the gentle slope that falls to the Bay.

—ELIAS OLAN JAMES
THE STORY OF CYRUS AND SUSAN MILLS

1-1

MAY FETE AT LAKE ALISO, 1917
Costumed students celebrated their May fete by becoming part of the campus' landscape of eucalyptus trees.

1-2, 1-3, and 1-4

COMMENCEMENT, CIRCA 1910
The Oval provided a dramatic ceremonial setting for Commencement.

1-5

MAY FETE AND EL CAMPANIL, CIRCA 1911
This May fete was celebrated next to El Campanil and under ancient oak trees.

The shared vision of 140 years sets forth a mandate for the future of Mills College to preserve, enhance, and build on this unique campus setting in the future.

INNOVATIVE EDUCATION FOR WOMEN

Mills College began as a much-needed educational facility for the daughters of pioneering families in the American West. Originally founded in 1852 as the Young Ladies' Seminary in Benicia, California, Mills was the first institution on the West Coast dedicated to educating women. Through a long arc from ladies' seminary to nationally-ranked college, Mills has held to its founding principles of innovative education for women. The approach to campus design reflects this philosophy and culture. Today, Mills College is an independent liberal arts college for women, with graduate programs for women and men, that educates students to think critically and communicate responsibly and effectively, to accept the challenges of their creative vision, and to acquire the knowledge and skills to understand the natural world and effect thoughtful changes in society. The curriculum combines traditional liberal arts with new educational initiatives that recognize the value of cultural, racial, and ethnic diversity. Forming connections among people, as well as across branches of learning, is fundamental to a Mills College education.

1-6

MAY FETE ON THE OVAL, 1947
May Day celebrations continued through the late 1940s.

CAMPUS PHYSICAL CONTEXT

The 135-acre campus is located in the hills above Oakland, California, very much in the physical center of the San Francisco Bay Area. The campus lies in the highly urbanized East Bay in the midst of predominantly residential neighborhoods. Interstate 580, completed in 1964, serves as the campus' northern border, separating it from Oakland neighborhoods. MacArthur Boulevard, a major arterial, forms the western and southern boundaries, with Seminary Avenue defining the campus'

eastern edge. The campus has had several entrances in the past, notably its original entrance off Seminary Avenue, that created a direct connection across Chimes Creek to the Oval and Mills Hall from Wetmore Gate, which was the major southern access to the campus. In response to shifting land use and public transportation, Richards Gate on MacArthur Boulevard now serves as the entry for the campus, beautifully highlighted by the London plane trees of Richards Road.

THE STUDY AND DESIGN PROCESS

Effective planning and design for Mills College will require an understanding of the resources of its campus cultural landscape, which have been identified through an assessment of its historical resources. The study began with the definition of project goals and research on past campus planning by the College. Subsequently, the study team undertook a cultural landscape survey and historic building documentation and developed a chronology of the campus' built environment. A cultural landscape survey identified the large- and small-scale features that create the distinctiveness of the campus. These include vistas, topography, building massing, focal points, courtyards, circulation, and tree canopy. The documentation of some thirty historic buildings and features follows criteria used by the National Register for Historic Places. Based on current planning for the campus and an understanding of its historic development and character-defining features, the College identified two locations as examples of how to integrate historic values into updating and expanding campus facilities, yielding conceptual designs for future implementation. The two studies addressed a range of program requirements to illustrate the flexibility and benefits of this process. As part of the educational component of the study, Mills students conducted research in the College's archives. Some prepared papers on a place of significance in the history of Mills, while others investigated how to conserve the campus' resources for a sustainable future. Four public lectures addressed the historical and cultural impact made by notable individuals, whose work shaped the campus landscape and architecture and formed its reputation as a progressive institution dedicated to women's education.

1-7

CAMPUS HEARTH ON PINE TOP HILL PICNIC, CIRCA 1960
Students and faculty often held picnics on Pine Top Hill enjoying the distant vistas of San Francisco Bay.

2.
EXPLORING THE HISTORY OF THE CAMPUS
HOW THE LANDSCAPE DEFINES THE COLLEGE'S IDENTITY

Since purchasing the land in 1868 in Oakland, California, the leaders and designers of Mills College have always embraced its natural character. Born out of the 19th-century American picturesque landscape era, which borrowed from English landscape gardening ideals and the natural order of landscapes, the early campus embodied a land design ethic that nurtured a progressive women's college.

My dear (Cyrus) had been a real landscape artist.

—SUSAN T. MILLS, CIRCA 1910

As you enter Mills College from the west through Richards Gate, a change in the visual atmosphere causes an involuntary reorganization of one's thoughts. You know you have entered a special place, a place remote from the outside world and one that has been in existence for a long while.

Further in, the composite character of the campus is eclectic, and the activity of the student body mirrors the boldness of the natural and built environment. The campus is the very definition of a mature California cultural landscape—one that has developed over time under the aegis of multiple talents and enlightened intents, whose ideals inevitably collided with the reality of programmatic challenges—the milieu of any prestigious yet aging campus.

These 135 acres of naturally undulating topography and ancient creek channels have, for more than a century, hosted a design imposition of exotic tree canopy and understory. Genus and species from every livable continent on the planet contribute to the contrivance. Truly, as a botanical metaphor of California itself, these staged forests became acculturated to their new environment.

Within the landscape, well-executed European revival architecture reflects nearly every design trend of the 19th and 20th centuries. Here are significant expressions of the French Second Empire style (Mills Hall and Sage Hall), the modest frontier folk Victorian (President's House), the Spanish colonial revival style with exquisite Andalusian cues (Music Building), and a democratic interpretation of beaux-arts classicism (Carnegie Hall, first known as Carnegie Library, and Wetmore Gate).

2-1

ENTRY TO MILLS COLLEGE, 1870s
Bridge over Chimes Creek at entry from Seminary Avenue.

What remains of the indigenous California oak woodland and once vast open meadows has become a well-photographed historic memory. Yet, despite the predominance of exotic species and the loss over time of open-space settings and select architectural gems, the campus continues to convey historic character from its initial period of significance—that of the heartfelt original vision of its founders, Cyrus and Susan Mills.

Rancho San Antonio

In northern California's pre-history, the land that is now Mills College was part of the ethnographic territory of coastal Native Americans. The tribes would annually burn off open meadows to ensure fresh herbaceous vegetation. This treatment of the land, common throughout California, was the first cultural alteration to the landscape, erasing any memory of a pristine natural ecosystem.

In the late 18th century and well into the 19th, under both the Spanish Colonial and Mexican Republic periods, the land was used for intensive ranching, causing further disturbance. Lands east of San Francisco Bay became Rancho San Antonio, some 44,800 acres granted to the military loyalist Don Luis Maria Peralta in 1820 prior to secularization by Spain. Peralta began his military career under Spanish rule at the Monterey Presidio and ended it with a long tenure at the Presidio in San Francisco. In a strategic political move, Peralta allied himself with the new Mexican government and was rewarded with a sizable land grant.

The boundary of the Rancho went from the San Francisco Bay east to the ridge tops encompassing the areas that are now Berkeley, Albany, Emeryville, Alameda, Oakland, Piedmont, and San Leandro. Rancho San Antonio was one of the more successful ranchos because of its proximity to the Bay, trading in food, hides, tallow, and construction materials, the mainstays of the period.

Rancho San Antonio was closest to Mission San Jose, located in what is now the city of Fremont. As the mission system declined in the 1830s, local resources were divided among the Peralta family and later many prominent pioneer families of Alta California. At its peak, the San Jose mission's assets were reported at 25,000 head of cattle; 19,000 sheep, goats, and pigs; 1,200 horses and mules; as well as a chapel, storehouses, workshops, mills, dormitories, orchards, vineyards; and a community of 2,500 men and their families. Historic structures associated with the Peralta family still exist in Oakland, San Jose, Fruitvale, and San Leandro.

Commemorating this period of history, the East Bay Chapter of the Daughters of the American Revolution (DAR) placed a plaque at Mills College on the northwest-facing summit of Prospect Hill in 1969, marking the spot where a Spanish Mission team of surveyors was known to have camped:

> 27 March 1772 Captain Don Pedro Fages, Fray Juan Crespi, fourteen soldiers, and a Christian Indian camped here and took a latitude sighting of the Golden Gate.

TWO PERIODS OF HISTORIC SIGNIFICANCE

Research and analysis of the campus' cultural landscape revealed two periods of historic significance that by definition contain works of regionally and nationally recognized landscape architects and architects whose designs embody the high artistic values of their times.

As depicted and discussed on the following pages, these two periods are:

- The Founders, 1868–1916
- The Modern Campus, 1916–1949

2-2

MILLS CAMPUS, 1916

At the end of its first period of historic significance, the campus had in place key iconic features such as Mills Hall, the Oval, El Campanil, Carnegie Library, Lake Aliso, the Bryant eucalyptus path, and extensive forestation.

2-3

MILLS CAMPUS, 1949

The western addition to the campus marks the second period of historic significance that ended in 1949. This addition includes many of the campus' notable features such as Richards Road with its composition of London plane trees, the Music Building and Greek Theatre, Orchard and Meadow halls, and the Art Gallery.

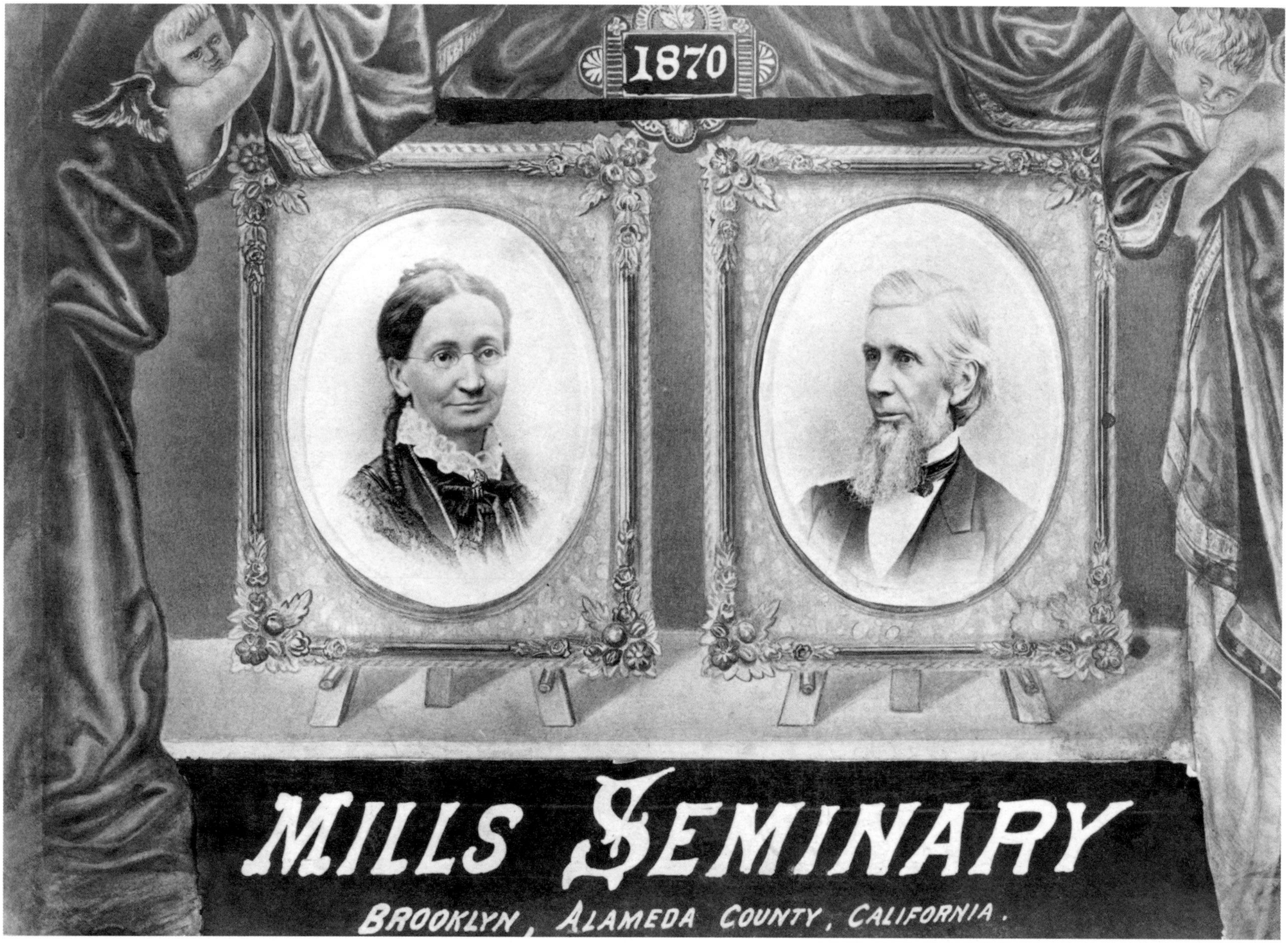
1870
MILLS SEMINARY
BROOKLYN, ALAMEDA COUNTY, CALIFORNIA.

FOUNDERS–THE FIRST PERIOD OF SIGNIFICANCE

1868–1916

Cyrus Taggart Mills (1819–1884), Susan Lincoln Tolman Mills (1825–1912)
The English Country Estate and the Picturesque Era
S.C. Bugbee & Son, Architects
Julia Morgan, Architect
Bakewell & Brown, Architects

Susan Lincoln Tolman graduated from Mount Holyoke Female Seminary in South Hadley, Massachusetts, in 1845. She remained at the seminary as a teacher inspired and encouraged by her mentor and renowned educator, Mary Lyon. Cyrus Mills, a graduate of Williams College in Williamstown, Massachusetts, with a doctorate in divinity, was a determined, faith-driven man inspired by his mentor, the noted educator Mark Hopkins.

Williams College, credited with being the progenitor of the American Foreign Mission movement a generation before Cyrus attended, trained young men to go into pagan countries and teach Christian ethics of civility and morality. Cyrus' dynamism presented itself early in his career. He was known to have requested mission assignments "where no one else dares to go." According to the seminal biography, *The Story of Cyrus and Susan Mills* by Elias Olan James, Susan met and married Cyrus in 1848, in part to fulfill the marriage requirement for foreign mission travel. The young couple immediately set sail for their assignment in Ceylon (Sri Lanka), where Dr. Cyrus Mills served as the president of Batticota College for six years, assisted by Susan as the school principal.

Cyrus and Susan taught classes in domestic skills to local girls and helped supervise several day schools. In 1854, the couple returned to Massachusetts, both in poor health. By 1860 they had fully recovered and accepted a new assignment to the Sandwich Islands (Hawai'i), where Cyrus became president of O'ahu College (Punahou School), founded by the Reverend Bingham and Congregational

The estate in Alameda County seemed to Dr. Mills ideal for carrying out his cherished hope, and soon he and his wife agreed to "finish their work" in the new surroundings and devote their modest fortune and the strength of their mature years to the development of a woman's college.

—**ROSILAND A. KEEP**
FOURSCORE AND TEN YEARS: A HISTORY OF MILLS COLLEGE

2-4

SUSAN. T. MILLS AND DR. CYRUS MILLS, 1870
Approximately one year before completion of Mills Hall.

missionaries in 1841. The school was designed to educate the children of missionaries. Susan Mills taught in the school, emphasizing calisthenics and the physical sciences, two proficiencies she developed at Mount Holyoke.

By the mid-1860s Cyrus and Susan returned to America. Historians believe that the couple came back either because of ill health again or because they did not have a controlling interest in O'ahu College. Most probably they wanted to establish their own mission school. They were committed to preaching the faith and determined to "civilize" foreign pagan societies. But they must also have heard about the menacing situation in the new state of California. An urgent countrywide rallying cry had arisen from the pulpits of New England urging all Protestant missionaries to respond.

In *Americans and the California Dream 1850–1915*, Kevin Starr, former California State Librarian, vividly depicts this dramatic emergence of barbarism in the young state of California and how the American missionary movement reordered itself to address the need:

> Amidst a frenzied pursuit of gold, amidst violence and a lack of care, Protestant missionaries struggled for a better California. . . . Not content to exhort at a distance, men of the cloth joined the exodus, determined to coax California into right paths. . . . California's ministers considered themselves protagonists in an era of apostolic challenge. No missionary to a pagan land, they felt, ever faced a more difficult call: to bring religion to the most profligate society America had yet produced (California), to cry forth Christ in a moral wilderness.

This appeal might have been the impetus that brought the couple back to America or which directed them once here to take on their next venture of faith. After touring the San Francisco Bay area and its surrounding settlements, they were drawn to the Young Ladies' Seminary in Benicia, California, operated by the pioneer educator

Mary Atkins. The couple bought the Young Ladies' Seminary in 1865 with the intent of educating and refining the daughters of the frontier families. Susan Mills looked to her alma mater, Mount Holyoke, for assistance in accomplishing their ambitious task, and she and Cyrus persuaded several teachers to move to California.

Ultimately concerned for the future success of the school, Cyrus kept a steady eye toward the institution's financial sustainability. Benicia was one of five cities in early California that briefly held the title of capitol, sharing the distinction with Monterey, San Jose, Vallejo, and lastly Sacramento by 1855. Although Benicia's time had come and gone (1853–1854), it managed to keep an aura of importance and growth potential.

But the long effort to complete a transcontinental railroad projected for 1868, located the western terminus in Oakland, California. Acting on this promise, Cyrus began to speculatively purchase land in and around Oakland and San Leandro. His prescient purchases paid off, enabling the couple to buy 60 acres and move the Benicia school to a new site in the hills two miles east of San Leandro by 1871.

The site had been the old Tompkins property, a tract of rolling hills that provided an open clearing framed by two creeks with sweeping vistas of the surrounding countryside and the San Francisco Bay. Cyrus and Susan wanted to position the seminary well outside the city of Oakland as a remote retreat. A new train station being built in San Leandro, west of the campus, would ensure access to the school. A once-nondescript dirt road that accessed a cluster of folk Victorian farmhouses along a Post Road was expanded and became Seminary Avenue, which served as the campus' eastern boundary and connected the tiny campus core to the village of San Leandro. In a few years the first small-gauge streetcar would connect Mills to Oakland along its northwestern boundary.

Cyrus' original vision for the landscape of the new seminary was that of an English country estate, a vision he and Susan carried with them from their New England

2-5

MILLS HALL, 1870s
Flanked by Leona and Chimes Creeks, the campus setting is a landscape denuded by years of ranching. Early traces of the Oval appear in the foreground of Mills Hall.

origins. The land, essentially a denuded blank canvas, had suffered from more than a century of ranching, yet riparian corridors of native California oaks, buckeyes, alders, willows, and associated native species survived. There were enough alders to inspire Cyrus to name the school "Alderwood." However, the first national publicity referred to the college as Mills Seminary College, and the name persisted.

Cyrus immediately began grand plans for a principal seminary building, one that would hold all academic, administrative, and living space for the students—the preferred style for women's educational institutions at the time. The building site was in the clearing between the two creeks, but not equidistant from them. The couple preferred a spot closer to Leona Creek so that they could hear the soothing sound of

water from a top-floor suite. To design this prominent multi-function building, Cyrus commissioned one of the leading architectural firms in the Bay Area, S. C. Bugbee & Son, designers of the first Oakland City Hall, whose notable clientele included Leland Stanford and Charles Crocker. According to Cyrus' request, Samuel Bugbee rendered a design modeled after the main building at Vassar College in the most popular style of the day, Second Empire, inspired by the 17th-century French architect François Mansart. Cyrus and Susan wanted their new endeavor to be regarded as the "Vassar of the Pacific Coast."

2-6

RUSTIC BRIDGE OVER LEONA CREEK, 1880s

Proximity to the creek allowed the soothing sound of its waters to be heard from the top floor suite of Mills Hall.

In a May 1871 issue of *California Spirit of the Times,* the fledgling seminary was admiringly described:

> The Institution just completed is one of the largest in the United States and by far the best in every essential on this coast. We doubt whether there is a building in the east for similar purposes that can excel it, not excepting the famous Vassar Female College at Poughkeepsie. The location . . . is more desirable, being free from fogs, well sheltered from the winds, and abounding in picturesque views. . . . The comfort of students has been consulted in every particular, and in fact not a single thing has been omitted deemed necessary for the happiness, progress and comfort of those for whom the great undertaking is intended.

Climbing to the top of the blue-domed cupola of Mills Hall the reporter continues his rhapsodic appraisal of what he encountered:

> The San Bruno hills rise in majesty, lights and shadows softly tinting their crest and sides; at their feet the waters of San Pablo Bay lie in sparkling beauty. The southern portion of the city, including Hunter's Point and the dry dock, is in view. Hundreds of little cottages and farmhouses dot the valleys on the Alameda side of the bay. To the left may be

> seen a portion of the Santa Clara Valley, the beautiful village of San Leandro and the fruitful fields and rich bearing crops of the thrifty tillers of the soil. To the right, San Francisco opens up in the distance. Alameda, with all her beauties of taste, skill, wealth, and Fruitvale, with its multitude of charming nooks and residences, form a lovely foreground, while old Tamalpais looms up as a sort of guardian of the vast array of natural beauties. The background is made up of hills, rolling ground and a distribution of attractive points. . . . The drive from the railway station is a beautiful and attractive highway. It is one mile in length and sixty feet in width with six hundred eucalyptus trees planted along its entire length. The entrance to the grounds is through a fine driveway thirty feet in width, under old oaks and over a fine bridge thrown across a ravine. There is to be a large lawn in front of the main building. A handsome chapel is to be erected about a quarter of a mile below, thus giving a pleasant walk. A fine gymnasium and observatory, summerhouses and rustic bridges are yet to be completed before the opening date in August.

Mid 19th-century American estate and campus design embraced the fashionable picturesque landscape style and Second Empire architectural styles. The picturesque style evolved out of the English landscape gardening school developed from a long pedigree in England, but adapted to America, or at least made nationally popular, by A. J. Downing (1815–1852) in his how-to book of 1841, *A Treatise on the Theory and Practice of Landscape Gardening, Adapted to North America*. The book addressed the arrangement of horticulture, the placement of architecture, curvilinear circulation, prominent entries, but most important, vistas of captured and composed scenery. Downing's basic tenet was that nature abhors a straight line, rejecting the strict geometry of French formalism and beaux-arts bilateral classicism. He taught the

2-7

MILLS HALL, 1880s
Cyrus' plantings begin to surround the campus buildings.

2-8

PROSPECT HILL, 1880s
An early vista from Mills Hall of Prospect Hill shows a land undeveloped.

2-9

MILLS SEMINARY AND CYRUS' FOREST, 1890s
Within 20 years, Cyrus' forestation efforts are well on their way to transforming the campus setting.

1203 – ROSE PORCH, MILLS COLLEGE, CALIFORNIA.

estate owner how to compose natural "pictures" and assured the novice that romanticism in literature could be applied to nature.

In *Four Score and Ten Years,* Rosalind Keep describes the progress of the grounds:

> The gardens were enlarged, orchards were planted, stables built, and acres of the campus set aside for kitchen gardens, grain fields, and pasture land for the large dairy herd. Thousands of trees were set out on the hillsides. Dr. Mills was interested in the many varieties of eucalyptus recently introduced from Australia, and his experiments with these foreign trees were watched with interest throughout the state. He delighted too, in many varieties of pine and poplar, acacia and elm, and procured and planted unusual foreign trees. Climbing roses were planted in profusion, and the white and yellow seedlings that were to attract visitors in the coming years to the "Rose Porch," the central porch of Mills Hall, were set out at this time. Purple wistaria [sic] vines began to climb their trellises to the third floor on the east side of Mills Hall.

In *California Gardens: Creating a New Eden,* David Streatfield offers a concise description of the California horticultural frontier during the Victorian era:

> Loudon's and Downing's styles were unthinkingly adopted in California, as elsewhere, because they were fashionable. Indeed, western garden makers were offered no other model to follow, since it was not until 1879 that Charles H. Shinn published *The Pacific Rural Handbook,* the first book to address specifically California's planting conditions. What distinguished the Californian interpretation of the different imported styles was the use of an extraordinarily wide range of plants. The state's first nursery was established in 1849 in Sacramento, and the rapid growth of it and other nurseries

2-10

POSTCARD OF MILLS HALL ROSE PORCH, CIRCA 1910

As described by R. L. Tabor in the College's 1885 viewbook, the "Rose Porch Entrance is so named from the Lady Banksia Rose, which completely hides the great fluted columns, and runs up on the building 45 feet. It is all from two roots and has several branches as large round as one's wrist. It blossoms early in May in immense clusters of snowy white, and is a sight which visitors come many miles to see."

2-11

EUCALYPTUS BRYANT PATH, 1880s
Three rows of eucalyptus comprised the path in its early years.

2-12

EUCALYPTUS BRYANT PATH, 1890s
Later in that decade as the plantings grew, the middle row of trees were removed.

2-13

EUCALYPTUS PATH, 1968
By the middle of the 20th century, Bryant Path had become a dominant element of the campus landscape and culture.

offering a multitude of nonnative plants in what was still a frontier region was a remarkable phenomenon. From the late 1840s through the 1920s nurseries dictated garden character in various ways, but it was the nurseries founded in San Francisco, Oakland, and San Jose, in the 1850s and 1860s that became the most important in the state, patronized both by urban gardeners and by owners of rural estates.

During the construction of Mills Hall, Cyrus began foresting the perimeters of the property, as well as the slopes, lanes, and a mile of Seminary Avenue mostly with the fast-growing eucalyptus. A double soldier-row of *Eucalyptus globulus*, blue gum, was planted along the western border, which eventually shaded the long path to the streetcar, and is presumed to be what was once the revered Bryant Path (now Kapiolani Road). Cyrus' favorite poet was William Cullen Bryant, and so a Bryant Literary Society was founded on campus. Cyrus planted other species—poplar, acacia, elm, olive, pine, magnolia, and other unusual foreign trees he collected from the couple's travels—placing them in meaningful ways in the landscape. He took his design cues from the robust horticultural scene that was well in place in the Bay Area.

In *American Space: The Centennial Years: 1865–1876*, John Brinckerhoff Jackson writes:

> Within a hundred mile radius of San Francisco a transplanted East had evolved. . . . So there could be no such thing as too much vegetation; more and more foreign plants appeared in gardens.
>
> There was botanizing, and to the parklike landscape of venerable live oaks growing on the sides of bright green hills the excursionists responded appropriately by citing Olmsted and Downing and Weidenmann. Whatever educated circles in the East thought and did, educated circles in the Bay Region were prompt to think and do. Tree planting was an Eastern fashion; eucalyptus trees suddenly became the rage in California, and an Oakland nursery sold one hundred thousand in a year—as windbreaks, as ornaments, as potential fuel.

Cyrus was not a professionally trained campus planner or landscape designer, nor was he knowledgeable about the characteristics of exotic plants. He was self-taught, intuitive, experimental, and driven by his vision for an aesthetically pleasing, permanent educational institution of moral character for young women of the West. It is not known if he used any of the popular books on landscape design, although a copy of an 1850 edition of Downing's treatise is in the College's Special Collections. Susan's younger sister Jane, who taught at Mills, had one of the finest private libraries on the West Coast, and Cyrus may have availed himself of these resources.

It seems clear, though, that the landscape design for Mills Hall was as sophisticated as the illustrations in Downing's book: a large grassy oval frontispiece the width of the front elevation of Mills Hall, a dramatic entry from Seminary Avenue over Chimes Creek on a romantic-period bridge, an almost ceremonial carriage-ride experience to the front door, and endless vistas. The Oval is the most central and oldest landscape design element on campus. Part landscape, part circulation, this green expanse

once held all formal and informal events, rituals, celebrations, dedications, and graduations. Though a novice in the field, Cyrus held his own with the highly trained designers of the day.

It also appears that Cyrus paid attention to what was being planted on the grounds of the University of California, Berkeley, and at the San Francisco Presidio. As with these landscapes, Mills was planted with large swaths of the wildly popular Australian blue gum, *Eucalyptus globulus*, in the early 1870s. The layout of Mills further mirrored the Berkeley campus to some extent, both sharing character-defining features such as great stretches of land with upward undulating topography, clearings at the top framed by oak-lined creek beds, and sweeping views of the surrounding countryside and San Francisco Bay. Both campuses also employed the Second Empire style of architecture. The Berkeley campus faced west, whereas Mills faced south; and Berkeley was a publicly funded land grant institution with four times the acreage, whereas Mills was a modestly funded private institution. Nevertheless, the new Mills Seminary was greatly respected, a frontier contender, fulfilling a cultural and academic need. It was known early on as the "castle among the hills" by Berkeley students.

In *The Story of Cyrus and Susan Mills*, Elias Olan James wrote of Susan's remembrance of her husband and his contribution to the campus grounds:

> She hoped the campus would never be spoiled. Forty years, now, since Cyrus had planned the landscaping, and the years had justified his imagination. Visitors driving up the Avenue turned in at the tree-bordered bridge, and there, in a breath, was the Oval, every tree, every shrub around it, a part of a balanced picture; and if the Rose Porch was in bloom! No wonder that alumnae and strangers alike exclaimed over that view. . . ."My dear" had been a real landscape artist.

2-14

MILLS COLLEGE, CIRCA 1920
Through the 1920s, the Oval and the buildings remained the academic and cultural center of the campus.

2-15

TOLMAN HOUSE, CIRCA 1893
Sister of Susan Mills and a member of the faculty, Jane Tolman resided in the Tolman House along with other faculty and visitors.

2-16

SAGE LIBRARY, CIRCA 1888
Adjacent to Mills Hall, the Sage Library was surrounded by gardens.

From 1870 until his death in 1884, Cyrus Mills planted 50,000 trees and shrubs, mostly from seed; constructed footpaths, bridle trails, winding drives for horse-drawn carriages; planted orchards; and directed the construction of all facilities in pursuit of a sustainable campus. During this time period, several support buildings were constructed behind Mills Hall, and extant structures were adaptively reused. Two new buildings were added, Tolman House and Sage Library, which contributed to the core campus that occupied the entire Toyon Meadow. While strategically planning for the College's future growth, Cyrus and Susan continued to purchase acreage on all sides of the campus. By 1900 the campus was more than 100 acres.

Mills College Seminary was incorporated in 1877; and with the appointment of its first Board of Trustees, the College was elevated to a more public profile. In their usual roles, Cyrus served as president and Susan as principal. Tragically, in April 1884 Cyrus died from complications of amputation surgery. After his death, Susan remained in her position as principal, declining an anticipated appointment to the presidency, while the board continued to shepherd the growth of the College. A new College Hall was erected for additional student housing, and in 1889 Mills crossed another threshold when it conferred its first baccalaureate degree.

Lake Aliso was dammed in 1888 for flood control and for irrigation water to ensure the beauty of the grounds. The lake, an ancient low spot created by an underlying earthquake fault and adjacency to Pine Top Hill summit, may have served as a watering hole for livestock during the Spanish/Mexican period. As part of the upstream watershed, the lake retained water in a wide basin and directly fed downstream Leona Creek that coursed the length of the campus. Lake Aliso later evolved into a primary venue of almost mythic stature for campus rituals. Senior pageants on the lake became a long-held campus tradition.

Susan Mills slowly recovered from her grief and in 1890 was elected president of Mills College. With a renewed sense of purpose and energy, she worked tirelessly to continue the couple's original vision by raising academic standards and securing recognition for the school. In 1890, the premiere historian Hubert Howe Bancroft noted in his multi-volume history of California: "Among private schools, Mills Seminary, now Mills College, an institution for the education of young women, has maintained a high reputation." By then, the campus grounds had matured and garnered a reputation for their parklike beauty.

2-17

LAKE ALISO, CIRCA 1916
Lantern procession and bonfire.

2-18

LAKE ALISO, 1920s
The lake provided an expansive and tranquil setting.

As president, Susan Mills took up where Cyrus left off. Her ambitious list of projects included the Nathaniel Gray Hall of Science, 1892; Lisser Hall, 1901; El Campanil, 1904; Margaret Carnegie Library, 1906; Kapiolani Cottage, 1909; the Gymnasium, 1910; and Alumnae Hall, 1916. Her first project, the Nathaniel Gray Hall of Science, was sited across Leona Creek from Mills Hall. It marked a change in course and curriculum for the women of Mills and exemplified one of her life-long interests. Paralleling a turn-of-the-century demand for natural and physical science education, Susan was at the forefront of a national trend. The building was made possible by the generosity of Nathaniel Gray, a pioneer and Gold Rush '49er. The architect, George A. Bordwell, designed the building as an unadorned, late Victorian, four-story, vertical mass in a creek setting with a rustic bridge linking it to the core campus. It provided generous laboratories, offices, consultation spaces, lecture rooms, and space for scientific display. Legend has it that the planting of ivy on the new building came from cuttings brought back to Mills by Harvard graduate and Mills Trustee, Warren Olney.

Susan Mills' next major project was inspired by a student initiative and the need for a general use auditorium. Again, the site was along the western bank of beautiful Leona Creek. Lisser Hall, named for the first director of the music department, Dr. Louis Lisser, was designed in a muted classical Greek revival style, with fluted Ionic columns, facing south as was Mills Hall. Designed by the San Francisco-based architects G. W. Percy and Willis Polk, Lisser Hall was completed in 1901 and served as a multifunctional campus facility used as a gathering hall, a chapel, an auditorium, and a music hall.

2-19

COLLEGE HALL, CIRCA 1890
College Hall provided additional housing for students.

2-20

NATHANIEL GRAY SCIENCE HALL, CIRCA 1905
Legend has it that ivy on the building came from cuttings brought back to Mills by Harvard graduate and Mills Trustee, Warren Olney.

2-21

LISSER HALL, CIRCA 1910
Original entrance designed in the Greek Revival style faced south as did Mills Hall.

Susan's most distinctive project soon followed—a college bell tower, the 72-foot tall, five-story El Campanil. The designer was the daughter of a close friend, the young engineer and architect Julia Morgan (1872–1957). El Campanil, which predated the Campanile at Berkeley, may have been Morgan's first commission after establishing her private practice around 1903. Morgan was the first woman to graduate from the University of California, Berkeley, with a degree in civil engineering. Following that achievement, Julia was encouraged by her mentor, the famous Bay Area architect Bernard Maybeck, as well as the noted New York architect Pierre Le Brun, a maternal cousin, to further her education in architecture at the Ecole des Beaux-Arts in Paris, their alma mater.

Originally the Académie des Beaux-Arts, the Ecole was founded in 1648 by Cardinal Mazarin to educate the most talented students in architecture, drawing, painting, sculpture, and other fine arts, with studies focused on classical arts and architecture from Ancient Greece and Rome. Under the control of the French monarchy, the school guaranteed a pool of artists for the noble classes. Louis XIV was known to select graduates from the school to decorate the royal apartments at Versailles. Napoléon III granted the school independence from the government in 1863, launching the French Second Empire era and the revitalization of Paris with grand boulevards lined with mansard-roofed buildings.

In 1896 an enthusiastic Morgan traveled to Paris and applied to the Ecole, only to be turned away. She stayed in Paris, continuing to reapply and retest (she tested 13th of 376) until she wore them down and was finally allowed admission. Julia Morgan had the distinction of being the first woman to be granted a Certificate of Arts from the Ecole. She returned to the Bay Area in 1902.

Morgan began her architectural career employed by John Galen Howard, another Ecole-trained architect, who was a distinguished professor and the designer of the University of California's classical beaux-arts ensemble. Her first project was to render

2-22

JULIA MORGAN, 1896
Julia Morgan designed many of the campus' memorable buildings during the first period of historic significance.

drawings of the Hearst Mining Building, Howard's first project on campus, funded by Phoebe Apperson Hearst in memory of her husband Senator George Hearst. Morgan partnered with Bernard Maybeck on the Greek Theatre at the University of California, Berkeley, as well.

Morgan designed El Campanil in a pure Mission revival style and nestled it among the ancient oaks near Chimes Creek facing Mills Hall. In the foreword to *California's Mission Revival,* by Karen Weitze, Harold Kirker wrote, "California architecture came of age with the Mission Revival. After more than a century during which successive immigrant groups imposed inherited building forms upon a land whose physical diversity encouraged cultural colonialism, the Californians discovered in the Franciscan missions a source for a distinctive regional style." The bell tower was funded by Mr. and Mrs. Francis Marion "Borax" Smith. F. M. Smith, known mostly for his mining interests, was also the Oakland streetcar entrepreneur who promoted consolidation of all urban rail into what was eventually known as the Key System.

Morgan used the newest technology of steel-reinforced concrete in structural details for the tower. The once-inoperable, homeless bells that had originally been cast for a California display at the 1893 Columbian Exposition in Chicago and later used at the 1894 Midwinter Fair in San Francisco finally found a permanent home in this structure. The spectacular El Campanil was completed in 1904 and celebrated with much fanfare and a formal dedication. Two years later the new tower survived the 1906 San Francisco earthquake with no damage. Following this structural success, Morgan was commissioned by the Mills Board of Trustees to assess existing buildings on campus for damages, and she subsequently filed a report with her recommendations. Morgan's construction prowess, demonstrated by the survival of

2-23

EL CAMPANIL DEDICATION CEREMONY, 1904

The Oval served as the gathering area for the dedication ceremony.

2-24

EL CAMPANIL, 1920s

Its high integration with the surrounding landscape gave this campus landmark much of its visual strength.

El Campanil, gave her almost instant regional credibility and drew commissions from a number of prominent clients.

Morgan's next Mills project was the Margaret Carnegie Library in 1906, named and funded by the steel king, Andrew Carnegie, as a tribute to his young daughter. This time Morgan experimented with an eclectic interpretation of Mediterranean revival with Italian Renaissance influences. Her classical beaux-arts training was fully expressed in this piece. With a perfectly symmetrical front facade and generously placed Palladian windows, red clay roof tiles, and balconies with stylized ornamental iron, the building resembled a villa outside of Rome. The second-floor Bender Room (1954 to present), with views of the Oval, once housed 7,600 volumes and stands today as an exquisite display of muscular redwood trusses, paneling, and built-in bookcases. A classical, double-tiered fountain with a sculptured ibis pedestal rests in the open lawn outside the library—an attraction to students and faculty for moments of repose. The introduction of the library in the formal center of the core campus elevated the academic and aesthetic feel of the campus landscape.

2-25

MARGARET CARNEGIE LIBRARY, CIRCA 1906

Designed by Julia Morgan, the library elevated the academic and aesthetic character of the campus landscape.

2-26

CARNEGIE FOUNTAIN, CIRCA 1906

The fountain remains today in its original location and is in need of restoration.

Morgan's angled siting of both El Campanil and Carnegie Library has always been inexplicable. Neither structure followed a rational line of symmetry, which seems antithetical to her classical training. It leads one to believe that she deferred to the elliptical nature of the Oval, ringing it with her best architectural designs. Possibly she was being responsive to her client, Susan Mills, a trait that would define Morgan throughout her career.

Kapiolani Cottage was the next project of Susan Mills and Julia Morgan. It was built in 1909 and named for the Kapiolani Society, a Mills College club of Hawai'ian women and the donors for its construction. Kapiolani Cottage memorialized Cyrus and Susan's missionary experience at O'ahu College in Hawai'i. Susan maintained strong ties with her Hawai'ian counterparts, in particular, the last monarch of Hawaii, Queen Liliuokalani, whose nieces attended Mills Seminary in the late 1870s. Originally overseen by a trained nurse, Kapiolani Cottage served as the College infirmary from 1914 to 1924. In 1926 it became a residence for missionary travelers and later a residence for caretakers. Designed in the arts and crafts style, which was a constant style in Julia's repertoire, the two-story, wood-frame building was set

2-27

KAPIOLANI COTTAGE, CIRCA 1910
Julia Morgan, architect.

2-28

GYMNASIUM,1920s
Designed by Julia Morgan in the Spanish colonial revival style.

2-29

HELLMAN POOL, LATE 1920s
Since demolished, the pool lay between the gymnasium and Alumnae Hall, all designed by Julia Morgan.

2-30

SUSAN MILLS WITH "TEDDY," CIRCA 1905
Susan's dog, Teddy, was a constant companion to her.

2-31

SUNNYSIDE, CIRCA 1890
Resting place and memorial for Cyrus and Susan Mills.

into an east-facing Prospect Hill slope with a rustic stone retaining wall amidst the maturing forest planted by Cyrus Mills.

Julia Morgan quickly followed with the gymnasium building in 1910. Her last project on campus, Alumnae Hall in 1916, was the first facility for student gatherings. The structure was later part of a U-shaped complex that included a gymnasium and an outdoor swimming pool (added in 1923). Morgan designed Alumnae Hall in the Spanish colonial revival style with fluted neo-classical columns at the entrance and on the east side of the building. The front gable end of the raised porch entrance faces Toyon Meadow. The since-demolished gymnasium was designed in the arts and crafts style with distinct Asian influences. Hellman Pool completed the composition and made it a popular gathering spot for active students.

Julia moved Mills into the future with her artistry and attention to cultural cues. She began the needed transition from Victoriana for a Mills that would be further modernized by President Aurelia Henry Reinhardt and the architect Walter Ratcliff, Jr. Morgan's body of work at Mills College paralleled the tenure of Susan Mills, although Alumnae Hall was not completed until a few years after Susan's death. By 1919 Morgan was hired by William Randolph Hearst, son of Phoebe Apperson Hearst, to begin design development on his epic La Cuesta Encantada, Hearst Castle in San Simeon, California.

Susan Mills retired in 1909. An *Oakland Tribune* article reporting on her retirement described her as, "rather small in stature, with the quick, alert glance and sunny manner. She moves lightly, and walks at a brisk pace. Her personality is felt immediately. Mrs. Mills has indeed lived to a beautiful age, and her interest in current events is no less intense now than in the beginning."

2-32

PROPOSED GENERAL PLAN OF MILLS COLLEGE, 1915

Prepared by C. W. Dickey & J. J. Donovan, architects, and Howard Gilkey, landscape architect.

Susan died in 1912. Mills faculty and students remembered her as warm-hearted and impulsive, a genius in inspiring those around her to do their best. Both she and Cyrus are buried in the southern portion of the campus at Sunnyside, a site they originally chose for building a cottage where they could retire.

After the death of Susan Mills, the Board of Trustees entered a time of flux that lasted for nearly four years. Without the two dynamic founders and with a growing student population, the Board reached out for professional help to assist the College during the transition. To their credit they consulted the architectural planning firm of C. W. Dickey and J. J. Donovan in 1915, as well as the landscape architect Howard Gilkey, to generate a long-range comprehensive plan for the entire campus, an exercise that had never been done before. Cyrus and Susan's method of planning had been a very personal one, responsive to the needs of both the students and the grounds.

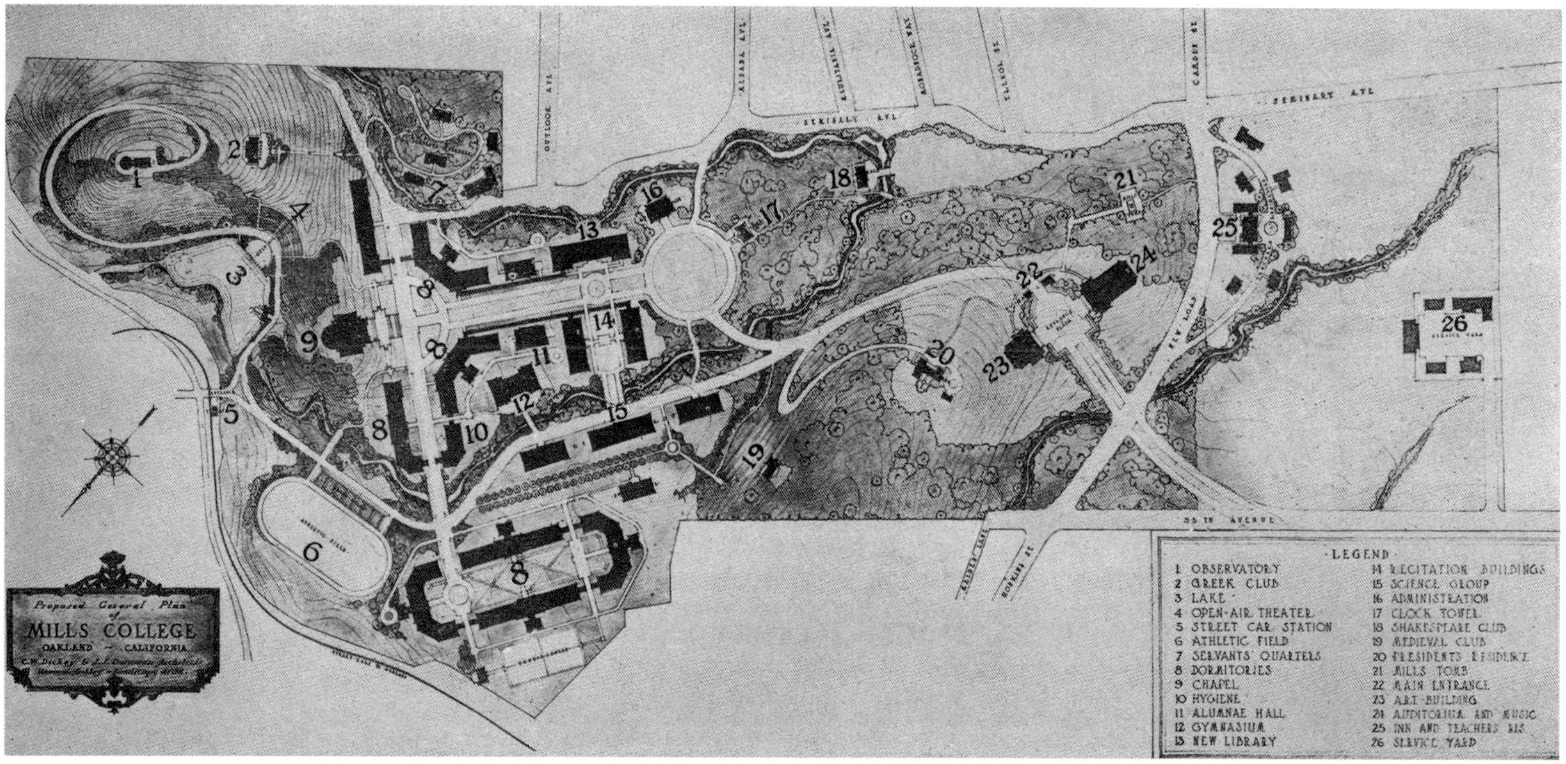

The efforts of the consultants resulted in resounding inaction, possibly because their plan required removing revered campus buildings, such as Mills Hall, and because its premise was to isolate the campus from the growing Oakland suburbs. The plan did, however, introduce the opportunities and wisdom of master planning.

In an expedient move the Board executed plans for much-needed student and faculty housing. The site for the new housing was the flat land of the upper west campus adjacent to Leona Creek, deliberately close to the California Railway Streetcar station. Signaling a shift to the west campus, construction of a suite of residential housing designed by the architects Bakewell & Brown (architects of the beaux-arts San Francisco City Hall complex and the Hall of Horticulture at the 1915 Pan-Pacific Exposition) continued in haste; Olney Hall was completed in 1917, Orchard Hall in 1919, and Meadow Hall in 1921. All were designed in a conservative yet eclectic style that captured nuances of institutional arts and crafts and hints of Italian Renaissance.

Olney Hall was named in honor of Warren Olney, who served on the Board of Trustees for 35 years. *The Story of Cyrus and Susan Mills* explains "[Susan] Mills had chosen Olney as her attorney after Cyrus' death, and she became the adoptive mother to the Olney family." Olney Hall's large entry court and garden area blended with the western bank of the sinuous Leona Creek. An enclosed courtyard was formed by the edifices of Olney and Orchard halls. It was designed by the landscape architect Howard Gilkey—a coincidence of sorts because Gilkey had teamed with Bakewell & Brown as the landscape architect for the Hall of Horticulture at the San Francisco Pan-Pacific Exposition.

The College seemed to be moving forward with or without a plan, and with or without a president.

2-33

ORCHARD AND MEADOW HALLS, 1920s

Viewed through the eucalyptus path from Kapiolani Road.

2-34

WARREN OLNEY HALL, CIRCA 1916

Designed in a conservative yet eclectic style that captured nuances of institutional arts and crafts and hints of Italian Renaissance.

2-35

OLNEY COURTYARD, 1972

Designed by Howard Gilkey, landscape architect.

Extant Historic Architecture

1 Mills Hall
2 El Campanil
3 Kapiolani Cottage
4 Sage Hall
5 Carnegie Library
6 Spring Cottage
7 Lisser Hall
8 Ross House and Larsen House

Landscape

1 Evergreen forests
2 Open field
3 Lawn
4 Parterre garden at Sage Hall
5 Orchards
6 Pastures
7 Kitchen gardens
8 Livestock pens
9 Other open spaces
10 Sunnyside
11 Lake Aliso
12 Creeks
13 Undeveloped

Summits

1 Pine Top Hill – 310′
2 Prospect Hill – 210′
3 Sunnyside Hill – 135′

Entries

1 Seminary Avenue entry to the Oval
2 Old Post Road entry

Circulation

1 The Oval
2 Seminary Avenue
3 Post Road
4 Kapiolani walking path
5 Surrounding roads
6 Streetcar line and Leona Station
7 Path to streetcar station
8 Bridges

2-36

MILLS COLLEGE, 1916
Elements of historic interest at the end of the campus' first period of historic significance.

2-37

AURELIA HENRY REINHARDT, CIRCA 1938
Mills College President from 1916 to 1943.

THE MODERN CAMPUS–THE SECOND PERIOD OF SIGNIFICANCE

1916–1949

Dr. Aurelia Reinhardt (1877–1948)
Walter Ratcliff, Jr. (1881–1973), Architect
The Spanish Colonial Revival–Eclectic Style
Howard McMinn (1891–1963), Botanist, Scholar
Howard Gilkey (1890–1972), Landscape Architect

I was the first college president who arrived on campus pushing a perambulator.

—**AURELIA HENRY REINHARDT**

At 39, widowed and the mother of two small boys, Aurelia Henry Reinhardt was not the typical choice for a college president, yet she had many friends on the Board of Trustees. She also carried an academic pedigree unusual for a woman at that time, especially in the West. She was the first woman to have a doctorate from Yale University, and the first woman to be granted an honorary degree from the University of California, Berkeley. Aurelia Reinhardt was also well connected in progressive, academic, and social settings in the Bay Area. She was a strong, confident woman of extraordinary potential.

It was the Board Trustee Ethel Moore who formally nominated Reinhardt for the position and ultimately made the motion to approve her appointment. The two women were friends and served together in several organizations, such as the Association of College Alumnae and the Equal Suffrage Association.

Although Dr. Reinhardt's confirmation was seemingly secure, qualms about her long-time affiliation with the Unitarian Church persisted. The Board felt compelled to first pass a resolution affirming that the College would remain "Christian" and not sectarian, and that proper and usual devotional services would be maintained. After that safeguard was in place, the Trustees cast their vote to appoint Dr. Reinhardt as president of Mills College on April 20, 1916:

> We have chosen Aurelia Henry Reinhardt to be president of Mills College because she sees in Mills College not a temporary job, a step to something better, but . . . considers

> it a challenge worthy of her steel, one which will engage her interest and her great gifts for all time. Respecting the past of Mills College and its present as the first woman's college of the West, she pictures for it a glorious future to which her labors will contribute.
>
> —Ethel Moore, Mills College Board of Trustees

Dr. Reinhardt made clear at her inaugural dinner in September 1916, as the *Oakland Tribune* noted, that one of her primary goals was "to keep the wholesomeness and beauty of rural life in our gardens along Leona hillsides." Indeed her sincere concern was for the grounds. "One of President Reinhardt's earliest recommendations to the Trustees was that a plan for architectural and landscape development of the College campus be designed before any permanent building be undertaken." She immediately began laying the groundwork for another master plan.

Dr. Reinhardt's career at Mills spanned from 1916 to 1943. At the beginning only eleven buildings were present. When she retired, the twenty-eight new and remodeled buildings attributed to Dr. Reinhardt's tenure included:

Warren Olney Residence Hall
Orchard House/Meadow House
Mary Kezar Chemical Laboratories
Community House
Toyon Hall
Steam Plant and Laundry
Beulah Gate and Lodge
Mary Camden Wetmore Gate, Lodge, and Pool
The Art Gallery
Ethel Moore Residence Hall
Hellman Swimming Pool

Music Building
Lisser Hall (remodel)
Norman Bridge Health Center
Mary Morse Residence Hall
Graduate House (remodel)
stables and riding ring
Grace Camden Richards Gate
Administration Building (a remodel of Sage Hall)
faculty homes and apartments

Mills College was fortunate to have Dr. Reinhardt, another dynamic and dedicated leader, to push the growth and success of the College, as the founders had done. However, the change from an 1880s campus to one of the 1920s-1930s was striking. To accommodate the automobile, unpaved carriage roads over rustic wooden bridges gave way to concrete infrastructure and paved roads. The landscape was no longer valued for pastoral visual components but for the ease of traversing through it briskly. Yet the new president was dedicated to retaining and enhancing the look of the campus.

Dr. Reinhardt was unwavering in her appeals to Bay Area society to assist in her goals. She was able to persuade Phoebe Apperson Hearst, the first female regent of the University of California system, to become a Trustee on the Mills Board, furthering Mrs. Hearst's interest in both women's institutions and education. It soon became apparent that when Mrs. Hearst became involved in projects, her preferred architect, Bernard Maybeck, was not far behind.

Bernard Maybeck, trained at the Ecole des Beaux-Arts in Paris, arrived in Berkeley in 1894. Known for his mastery of the arts and crafts and neo-classical beaux arts architectural styles, he was awarded the American Institute of Architects (AIA) Gold Medal in 1951 in recognition of his significant body of work and lasting influence on the theory and practice of architecture.

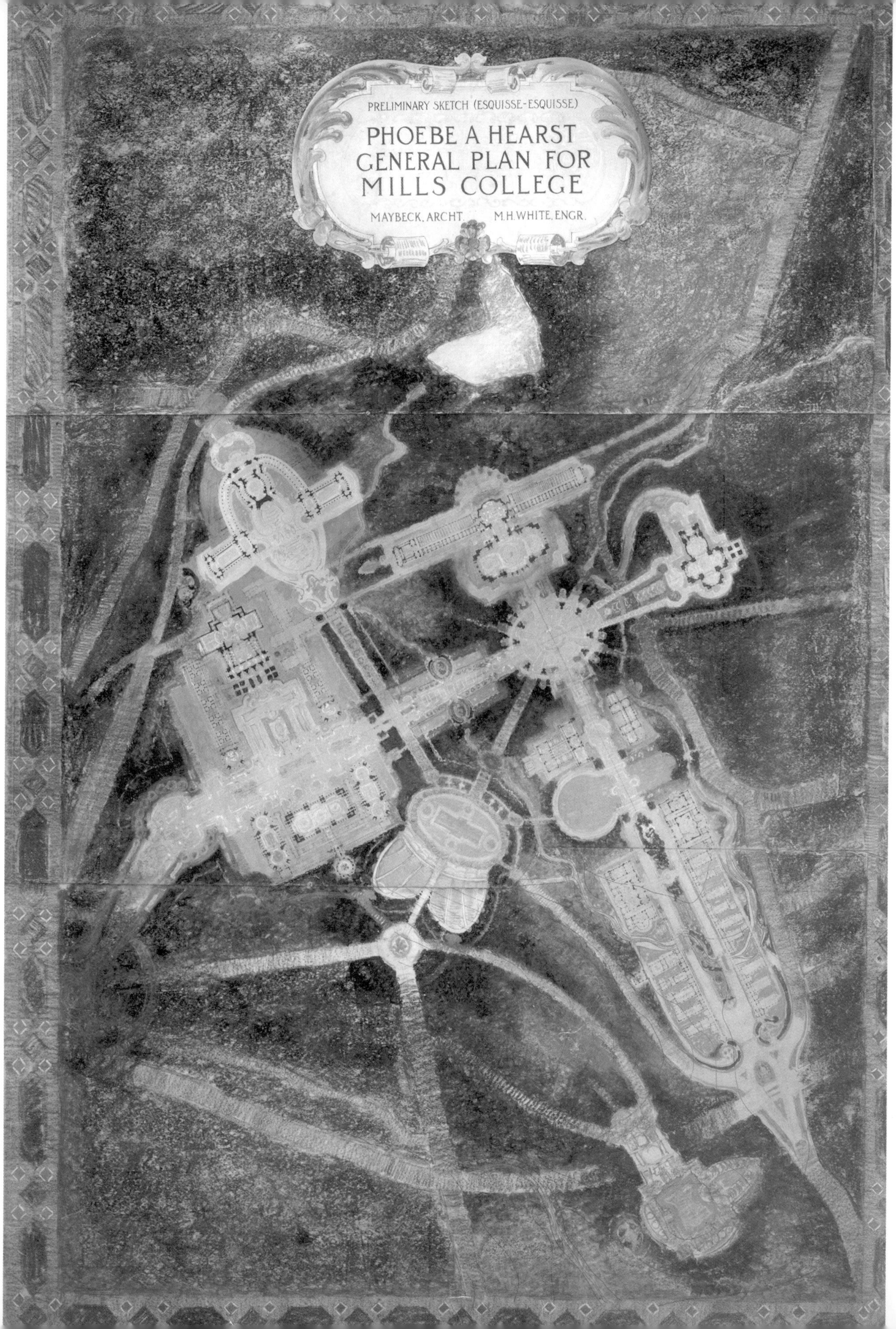
PRELIMINARY SKETCH (ESQUISSE-ESQUISSE)
PHOEBE A HEARST
GENERAL PLAN FOR
MILLS COLLEGE
MAYBECK, ARCHT.
M.H.WHITE, ENGR.

In 1918, with a student population expected to double, Maybeck was commissioned by Hearst, at the request of President Reinhardt, to develop a comprehensive campus master plan. Maybeck later described his plan as, "a subtle guide, leading the unsuspecting girls of Mills to hunger for ideals, like fine music stimulating to great deeds." Primary goals of the plan were to connect the campus with the Oakland community and disperse buildings into zones according to campus uses, based on two major axes.

Maybeck proposed that public buildings (the music auditorium, theatre, art gallery, and a complex for physical education) be sited close to the main gates, science laboratories on a road coming from the commercial district (Seminary Avenue), and a library at the heart of the campus. His plan, based on solid beaux arts principles, consisted of major buildings balancing each other across defined open spaces, dispersed throughout a generously landscaped campus. The plan combined classical yet modern design with symmetry. But it was dependent on a canopy of trees to tame the massing of the structures.

The plan deferred to the existing landscape, softening Maybeck's beaux arts allegiance. But, as the previous C. W. Dickey & J. J. Donovan plan demonstrated, his plan also anticipated the removal of the original core campus. He told the Trustees, "An Architectural Plan is not made for the glorification of the architect, but as a solution to certain problems . . . admitting flexibility as a fixed principle. . . . Each move in the development must add to the beauty of the whole."

Maybeck urged the Board to acquire a ten-acre parcel on the west side for a new entry and road and to acquire five acres adjacent to Lake Aliso, stating, "Lake Aliso should remain unchanged other than to accentuate the spirit that it now expresses, a solemn wilderness, and should remain as the holiest of holies." Maybeck's unrelenting advisement for these land acquisitions turned out to be his primary legacy, because his grand plan was never fully realized.

2-38

PHOEBE A. HEARST GENERAL PLAN FOR MILLS COLLEGE, 1918–1919

Preliminary sketch prepared by Bernard Maybeck, architect, and M. H. White, engineer.

2-39

PRESIDENT'S HOUSE, 1924
The old Spring Cottage was redesigned and enlarged to serve as the President's House.

Maybeck's concluding appeal to the Board was:

> The styles of the various departments should be established by all who are concerned in the welfare of Mills College and should not be left to the will of any one individual, for unless a tradition be established concerning the artistic expression that Mills should have, Mills College will grow into a hodgepodge of architectural styles like a graveyard full of monuments, to the glory of many an architect. This, however, is not the spirit in which Mills will work out the problem of building her future. Having once begun to beautify her campus, and having had the vision of possibility, Mills will safeguard its beauty.

Although many of Maybeck's ideas were inspired and tempting, the Board of Trustees was overwhelmed by his initial vision for an all new campus. Yet, the last iteration of his plan was adopted by the Board, although the logistics of demolition, construction costs, and phasing had not been fully investigated. Despite the Trustees' enthusiasm, Maybeck's 1918–1919 *Phoebe A. Hearst General Plan for Mills College* remained an unimplemented vision, but for unexpected reasons. Soon after adoption of the plan, Mrs. Hearst died. In her absence the arranged relationship between Mrs. Hearst's architect and the Board as well as the potential for a generous donation to implement the plan collapsed.

Such a setback at the beginning of one's tenure would have daunted a lesser person. But Aurelia Henry Reinhardt continued her campus-planning efforts. She scaled back her expectations and took a more moderate course. She asked the architects Bakewell & Brown to redesign and enlarge the old Spring Cottage, a small folk Victorian building (circa 1860) remaining from the Tompkins property period, and make it the President's House.

When Dr. Reinhardt signed on in 1916, she inherited a landscape that had matured nearly a half century from the early plantings of Cyrus Mills. Indeed, the campus was known for its beauty even if the buildings were showing a little wear and becoming dated. The landscape architect Howard Gilkey was already associated with the campus and Reinhardt supplemented his work with that of another landscape expert, the botanist and scholar Howard McMinn. It was unusual for a small private college to have two expert landscape professionals, both graduates of Professor John Gregg's Landscape Gardening and Floriculture Program at the University of California, Berkeley.

2-40

HOWARD MCMINN, CIRCA 1940

Howard McMinn promoted the planting of California native plants on campus.

Dr. Reinhardt encouraged Howard McMinn's interest in experimenting with California native plant material on the campus. The campus was losing some of its native California character, and McMinn was in a position to reclaim it and reverse the trend toward exotics. He and his students began collecting and planting California native stock to test conditions in the San Francisco Bay Area for purposes of ornamental use. In 1919 McMinn published *A Manual of Trees, Shrubs, and Vines of Mills College Campus*. McMinn promoted the planting of California native trees and shrubs—redwood, oak, manzanita, toyon, and ceanothus—throughout his tenure.

In 1921 Mills College bought the stately Victorian-era Kimball House (circa 1880) from the Morse family for use as a music building. Originally situated off campus further south on Seminary Avenue, it was moved to its present site at the confluence of Chimes and Leona creeks. The house was purchased by John Kimball in the late 19th century so that his daughters could easily attend Mills. The Morse family acquired the house at an undetermined date.

By 1922, Dr. Reinhardt began a search for an architect to design a plan for the campus that could be implemented. With her ties to Berkeley's social circles, she found the accomplished, well-respected architect, Walter H. Ratcliff, Jr., an ambitious, self-taught man who apprenticed under John Galen Howard. Ratcliff designed upscale houses as speculative ventures. His design styles were predominantly arts and crafts and European revivals.

In an oral history published in the Fall 2005 issue of the *Mills Quarterly,* Walter Ratcliff's son Peter notes a turning point in his father's career:

> I recall one Sunday we met (he took us along) Aurelia Reinhardt at the Oak Knoll Country Club. They had just completed their country club, which was a rather nice colonial Spanish. She said 'This is the kind of thing I want.' Dad and Mom packed up and went to Mexico. He took lots of pictures and did a lot of sketching.

Walter H. Ratcliff, Jr., taught himself the preferred style, choosing to learn the style from its source.

As the United States celebrated the opening of the Panama Canal in 1915 with a lavish San Francisco exposition, San Diego hosted a coincident event similarly meant to celebrate the linking of the Atlantic and Pacific oceans. This exposition created one of the most character-defining complexes of Spanish colonial revival architecture in the United States. This revival style and its many variations became the latest statement in architecture.

In his book *Architecture of the Southwest,* T. E. Sanford wrote:

> The marriage of what was seen as historic and logical appropriateness and unfamiliarity was a fruitful one . . . and Balboa Park became the birthplace of that Spanish Colonial revival which by 1925 had become a nationwide craze.

The master architect Bertram Goodhue was the creator of the San Diego Exposition. Walter Ratcliff and his wife may have passed through San Diego on their reconnaissance mission to Mexico.

2-41

WALTER H. RATCLIFF, JR., CIRCA 1925
Walter H. Ratcliff, Jr., reinvented Maybeck's plan for Mills College and introduced the modern era Spanish colonial revival architectural style to existing and numerous new campus buildings.

When Ratcliff returned, campus planning and construction began in earnest. Dr. Reinhardt made Ratcliff the official campus architect; and Howard Gilkey, the campus landscape architect. Aurelia had her design team in place by 1923, and she requested that the Board of Trustees codify the new architectural style as policy. Ratcliff's first accomplishment was to reinvent the Maybeck plan, an ongoing process that continued through 1930. Ratcliff was charged with modifying the work of a renowned architect and generating a plan that was physically and economically feasible. He commissioned a civil engineering survey, assisted by Gilkey, to create an exact base map from which to begin his work. Respecting the premises of Maybeck's work, Ratcliff sought to retain as much symmetry and functionality as possible, while adding his own imprint to the plan.

In a 1924 iteration of his plan, Ratcliff wrote on the obvious conundrum of working on an existing campus: "In the problem of making an architectural plan for an established college the considerations are threefold, a scheme must be established that takes into account: First the present buildings and grounds, second the necessary development in the immediate future, and third the development in the distant future, which is the ultimate scheme."

It is notable that in the campus plans generated over the years, each one retained Julia Morgan's El Campanil in situ. Conversely, those same plans called for the removal of the mother building, Mills Hall. It is not clear whether the Board of Trustees or President Reinhardt was hesitant to proceed with such a near-blasphemous act, or if budget concerns or politics prevented it. The fact that Mills Hall is standing today testifies to a reverence for the College's founders.

Walter H. Ratcliff, Jr., Campus Architect 1923-1947, completed projects:

- Beulah Lodge and Gate, 1925
- Art Gallery Building, 1925
- Ethel Moore Hall, 1926

PRO CHRISTO ET MUNDO
MILLS
COLLEGE
PLAN
W.H. RATCLIFF - JR.
ARCHITECT

- Wetmore Gate, Lodge, and Pool, 1926
- Lisser Hall remodel/reorientation, 1927
- Music Building and Greek Theatre, 1928
- Faculty Village (most residences), 1928-1940
- Orchard–Meadow remodel, 1929
- Norman Bridge Health Center (now the Vera M. Long Building), 1930
- Pine Top Hill Campus Hearth, 1930
- Richards Gate, 1934
- Mary Morse Hall, 1935

In 1925 the first design project for Ratcliff was to complete a new, western-oriented campus entry. On the newly acquired property, Ratcliff designed the simple, classical Beulah Gate and attached Lodge. The east/west Beulah Road axis was derived from Maybeck's plan, with construction and planting of the roadway already in process. Named for Beulah Brook, the smallest rivulet on the property, this fresh orientation signaled a strong connection to Oakland's emerging suburbs and future access to urban rail—the Beulah streetcar station was later designed for Mills College. Beulah

2-42

MILLS COLLEGE PLAN, 1930
One of several campus plans prepared by architect Walter H. Ratcliff, Jr.

2-43

BEULAH GATE, CIRCA 1930
Beulah Gate marked the new western-oriented entry for the campus created to meet the newly expanded streetcar system of the East Bay.

2-44

RICHARDS ROAD, 1930s
Lined by double-row plantings of London plane trees, Richards Road is one of the most superbly designed features of the campus landscape.

Road was the most important element in the modern campus design; and, as Maybeck had intended, it was to be a promenade in true European fashion.

One of the most superbly designed features in the landscape of today's campus is this entry road, now called Richards Road, inspired by the Ecole des Beaux Arts-trained Bernard Maybeck and assisted by the expert botanist Howard McMinn. The stunning promenade is lined by London plane trees, *Platanus x acerifolia*, assumed to be a hybrid between the American plane *Platanus occidentalis*, and the Oriental plane, *Platanus orientalis*. The planting scheme consists of widely planted double rows on each side of the road with walks between the rows. The tree planting layout is triangulated

to maximize the trees' broad-headed, spherical growth. Early pruning took place to ensure uniformity of spread and to increase the diameter of the canopies.
The choice of a deciduous tree was a remarkable departure from the generally monochromatic evergreen campus landscape. The collision of deciduous broad-headed plane trees against a backdrop of tall, vertical blue gums, *Eucalyptus globulus*, along Kapiolani Road symbolized where the old campus met the new. Whether intended or not by the designers, the planting evokes ancient symbolism—evergreen trees representing the eternal and the male of the human species, deciduous and fruiting trees representing both the cycles of life and the female.

The Art Gallery was Ratcliff's second project in 1925, and its design addressed access to the urban rail as well. The art building complex was made possible by the vision and initial philanthropy of Jane Tolman, sister of Susan Mills. The chair of the Art Department, Roi Partridge wanted "The finest, first-class and most permanent building upon the campus to be the heart about which will cluster ultimately studios and workshops for the crafts." The choice of site required relocating and again

2-45

ART GALLERY RENDERING, CIRCA 1924

Design study of tower for the Art Gallery prepared by Walter H. Ratcliff, Jr., architect

2-46

ART GALLERY, 1982

The south entry replaced the original north entry that was oriented to the streetcar line.

enlarging the President's House (Bakewell & Brown). This time it was moved to its present site west of Orchard Hall.

The original entry to the Art Gallery was on the north facade facing the streetcar line. From the station, a visitor was presented with a set of brick terraces descending down into the main building entry. Designed in the Spanish colonial revival style, the building's south-facing asymmetrical facade is composed of a central tower with two gabled tile roof forms splaying out at obtuse angles. The main gallery is housed in a rectangular building featuring an enormous skylight that fills the space with natural light. Opposite the main entrance, an elegantly proportioned, covered arcade connects to a lecture hall and classrooms. Ratcliff originally designed the Art Gallery to include a tower with an ornate top; but it was never completed, thus leaving

2-47

WETMORE GATE, 1930s
Wetmore Gate met Seminary Avenue north of MacArthur Boulevard.

the structure incongruously austere for the style. Two large, stone-carved fu-dog statues, a gift of Chinese art from Albert M. Bender, adorn the main entry steps to the complex. An allée of Lombardy poplar trees, *Populus nigra Italica*, planted in the exterior corridor, define the pathway toward the upper reach of Leona Creek.

In 1926, Maybeck's primary north/south axis was realized and became a double-lane vehicular road connecting the Oval to a new south entry, Wetmore Gate. The lanes were separated by a generously landscaped median that centered El Campanil at the north end and provided a stunning new classical entry at the south end consisting of an array of columns with ornamental iron gates, a gate lodge, and a large reflecting pool. The new south entry was named for the generosity of alumna Mary Camden Wetmore.

The first residence hall constructed from the Ratcliff plan was in honor of the student housing advocate and Trustee, Ethel Moore. A shield over the front entry is inscribed "Vassar and Mills" (Moore was a Vassar graduate). This rambling, faux-Spanish village has sixteen elevation changes, five patios, a beamed-ceiling common room, and a dining room. It is brilliantly set into the south-facing brow of Prospect Hill, "fitting the complexities of the surface of the ground." Although Oakland was building toward Mills on the west and north, maintaining a south-facing campus orientation was important to Ratcliff's plan. Landscape architect Howard Gilkey masterfully designed the interior courtyards and surrounding landscape, including a series of themed courtyards with different design narratives and species of plants. One such courtyard was planted with rhododendrons, pines, and olives that at just the right time, produced waves of fragrance to coincide with Commencement.

In Gilkey's narrative description of courtyard garden spaces, color was almost as important to him as species selection. His planting palette seemed nearly unlimited, displaying a variety of colors such as flame scarlet, copper, henna orange, soft buff yellows, and crimson. Gilkey further detailed the central area as "a motif of rolling

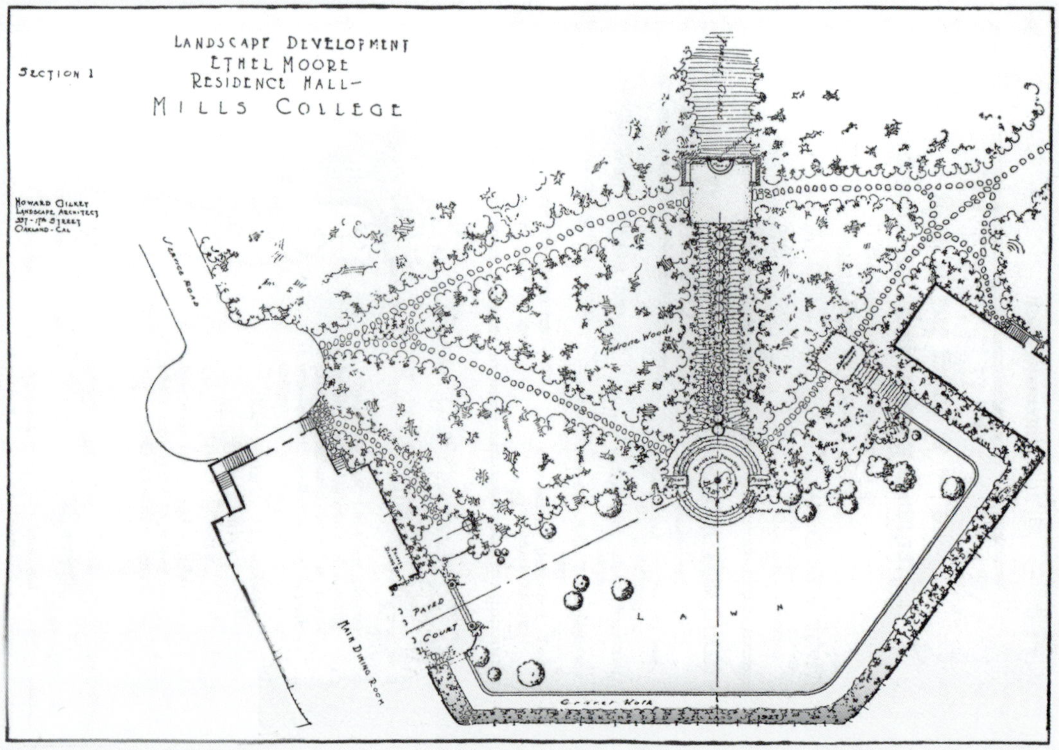
LANDSCAPE DEVELOPMENT
ETHEL MOORE
RESIDENCE HALL—
MILLS COLLEGE
SECTION 1
HOWARD GILKEY
LANDSCAPE ARCHITECT
OAKLAND - CAL
SERVICE ROAD
Madrone Woods
PAVED
COURT
MAIN DINING ROOM
L A W N
Gravel Walk

lawn . . . with broad circular stone steps. A fountain centering on a vista line from the dining room is the dominating feature of the dais as well as the court itself. It is the termination of a series of 27 circular fountain bowls arranged as a cascade."

Gilkey promises to achieve the finest landscape design for the College in the October 1926 issue of the *Mills Quarterly*:

> Ethel Moore is a case in point, and what we do for the gardens will undoubtedly become a standard which we shall try to excel in our future work. By and large, the distinction between environment and education is vague. Certainly, one of the distinctive things that Mills offers its students is the beauty of its environment, afforded and exemplified happily in the charm of its varied gardens. I feel that the destiny of the institution is involved in creating at Mills landscape gardening of the highest type possible.

In 1927 Walter Ratcliff was asked to remodel the Greek revival Lisser Hall to a design more compatible with the newly adopted Spanish colonial revival style of the campus. He not only remodeled the exterior but reversed all interior functions, creating a newly constructed classically columned north-facing front entry, oriented to the axis of the newly constructed Beulah Road. Three rounded arches open onto a full-height porch with three sets of double doors that, at the time, looked out on an open meadow and low-rise stairs running the distance of the elevation. A hipped tile roof completes this simple but elegant facade.

Of Ratcliff's work at Mills, the Music Building is generally thought to be his finest piece, garnering him a Northern California Chapter of the American Institute of Architects award for design excellence. Completed in 1928, this 31,000-square-foot two-story building is sited at the north base of Prospect Hill on the south side of what was then Beulah Road. The building addresses the linearity of the road with a long, asymmetrical, side-gabled front elevation and, typical for the style, holds

2-48

ETHEL MOORE HALL, 1933
The first residence hall constructed from the Ratcliff master plan.

2-49

COURTYARD FOUNTAIN, 1930s
The fountain reflects Howard Gilkey's attention to a detailed design aesthetic.

2-50

LANDSCAPE DEVELOPMENT FOR ETHEL MOORE HALL, 1925
Prepared by Howard Gilkey, landscape architect.

2-51

LISSER HALL REMODELED, 1950s
Walter Ratcliff redesigned Lisser Hall in 1927 to be compatible with the chosen campus architectural style and to reorient the building's entry to engage the newly constructed Beulah Road.

2-52

GREEK THEATRE, 1920s
This austere gathering space is part of the hillside that adjoins the south-facing facade of the Music Building.

2-53

HOWARD MCMINN'S EXPERIMENTAL GARDEN, 1930s
Directly across Beulah Road from the Music Building, this half-acre plot of land provided a place for McMinn's experimental plantings.

eleven levels. Designed in Ratcliff's most ornate interpretation of the Spanish colonial revival style, the building has Andalusian-inspired walled courtyards with Moorish tile fountains. A wide, Carmel-stone central pathway leads to the main entry with an exquisitely carved tympanum arch featuring classical statues of young women engaged in musical activities.

The adjoining Greek Theatre set further into the hillside is an austere yet impressive outdoor gathering space. The vista from the top rows features the Oakland hilltops in the distance. Prospect Hill beyond the Music Building has an approximately 70-year-old oak woodland planted by Howard McMinn. At first glance it seems to be naturally occurring.

Howard Gilkey designed the landscape along the entire front elevation of the Music Building, the front courtyard, and the adjacent naturalistic pond. In detailed drawings

Gilkey embellished the landscape design for the Music Building to match the design intensity of Ratcliff's architectural design. Landscape historian and Getty Grant lecturer Phoebe Cutler explained, "The grandeur gave Gilkey the scope to display some of his training in classic European gardens and his exposure to one or more world's fair. We see his fondness for axial perspective, as in the 44-foot-long, stone-laid path to a statue of Apollo; his weakness for water features, as in the pond and rivulet and a proposed Moorish fountain; and his delight in sculpture, as in Apollo, two dolphins, and two vases." A 1928 *Mills College Weekly* article also notes that "A great quantity of Carmel stone is being used both in the front and in the terrace and fountain constructions. Rhododendrons, tree ferns, and rose acacia have been planted in the front."

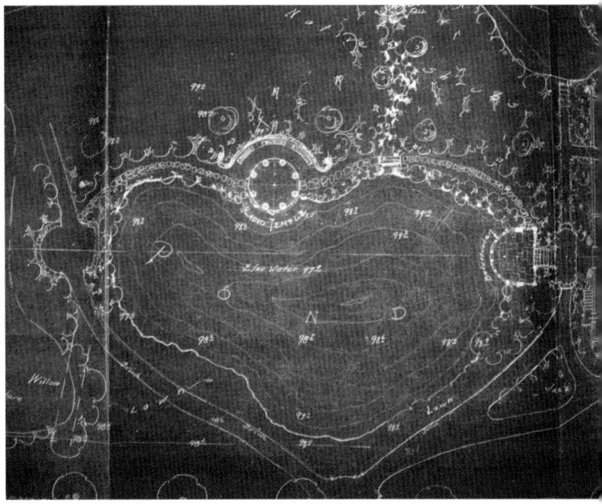

2-54

DESIGN FOR THE POND AT THE MUSIC BUILDING, 1928

Howard Gilkey's design for the pond adjacent to the Music Building was not fully realized as evidenced by the uncompleted "Temple to the Radio".

Across from the Music Building and west of Orchard and Meadow residence halls, Ratcliff allowed McMinn an approximately 75-foot by 300-foot swath of land for a botanical garden for his experimental collections. This area is now the President's Meadow. The gesture was most likely a place holder for an anticipated Ratcliff building that was designed and rendered but never built. For most of his career at Mills, McMinn had lobbied for a permanent experimental botanical collection area, but ultimately he was unsuccessful. However, over several decades in residence on the campus and between writing several scholarly reference books on California native plants, he planted more than 275 separate native species on campus. The following year he completed the first of three major test plantings of trees on the campus: 1,000 redwoods, *Sequoia species*; 250 Douglas fir, *Pseudotsuga species*; and 250 Sitka spruce, *Picea sitchensis*.

From 1928 to 1940 Walter Ratcliff, Jr., designed a group of nine single-family and duplex homes, known as Faculty Village, in the Spanish style for faculty on campus. More residences and gardens were added later. The original ensemble resembled a small subdivision or district. In the center of Faculty Village is an open green,

encircled by a road, similar to the Mills Hall Oval. Professor Howard McMinn and his family occupied the first unit.

In bringing existing buildings into design conformance with the new campus style, Ratcliff also made functional modifications to Orchard, Meadow, and Olney halls, joining them for common interior usage. The genteel Sage Library close to Mills Hall was made over as well, essentially subsumed within a Spanish stucco administration building.

By the late 1920s the campus landscape had matured beautifully, fully realizing the early vision of the founders. Under the supervision of Howard Gilkey and Howard McMinn, the grounds were in the care of experts, dedicated stewards of the horticultural collection. Howard Gilkey worked on landscape architecture projects hand in hand with the campus architect through the 1920s. In residence for nearly four decades, Howard McMinn had more opportunity to manage the grounds campuswide.

An article in the November 1928 issue of *Pacific Coast Architect* described the Mills landscape:

> The topography of the land is widely varied and includes rolling hills, small, level valleys, a small body of water, Lake Aliso, and two creeks which wind themselves through the grounds. For a number of years it can hardly be said that the landscaping progressed by any preconceived or definite plan, but notwithstanding this fact, no serious flaws of judgment or mishandling of the landscape are apparent. As a matter of fact, the campus is abundantly supplied with spots and vistas where the efforts of man and nature have been combined with pleasing effect.

2-55

ART GALLERY
The art gallery was originally designed by Walter Ratcliff, Jr., in 1925. This view is of the south facade.

2-56

MILLS COLLEGE CAMPUS, 1937
The heavy forestation of the original campus contrasts sharply with the newer plantings of the western addition to campus.

Walter Ratcliff continued his work with the Norman Bridge Health Center (now the Vera Long Building for the Social Sciences), completed in 1930. It was built in a former apple orchard north of Toyon Meadow above the old Post Road and accessed from Pine Top Road. He received another AIA award for design excellence for this building. The south-facing edifice ran for more than 100 feet and provided soleria and sunny porches, making it an ideal health center and clinic for the College. Three giant redwood trees planted by Cyrus Mills once stood in the front entry garden, which has since been replanted with California natives. Several oaks remain on the property including a Spanish cork oak, *Quercus suber*. The original north main entrance is surrounded by decorative glazed Spanish tiles and flanked by small, deep-set windows with wrought-iron grills, illuminating the nuances of the style.

In the mid 1930s the streetcar line serving Mills and the surrounding area was abandoned. The automobile had eclipsed mass transit. This gave the College the opportunity to purchase the remaining right of way and an additional parcel of land owned by the Presbyterian Church Board of Missions of New York. This last stretch of land presented the opportunity to construct a new gate along the main road, Hopkins Street, which was later realigned and renamed MacArthur Boulevard.

2-57

NORMAN BRIDGE HEALTH CENTER, 1930s
The south-facing edifice provided soleria and sunny porches.

2-58

GRADUATE HOUSE, 1930s
Ming Quong Home for Chinese Orphans. Designed by Julia Morgan and later acquired by Mills College.

The parcel owned by the church had been used for the Ming Quong Home for Chinese Orphans. The orphanage, later Alderwood Hall, and today the Julia Morgan School for Girls, was designed by architect Julia Morgan and dedicated in 1925. Her design was an eclectic mix of Spanish colonial revival and arts and crafts, with strong Asian motifs in the facade's terra cotta tiles, stucco walls, and roof details. The building and its attendant structure, Geranium Cottage, were used as the Graduate House for Mills until the 1950s.

Following the new land acquisition, Beulah Gate was dismantled in 1934 but the Lodge was left intact at the intersection of Beulah and Ming Quong roads. Walter Ratcliff designed a new gate of plastered concrete and stone curved walls with double square columns and attached pedestrian portals. The largest columns hold stylized lanterns—a gracious, welcoming entry. The new gate was funded by alumna Grace Camden Richards, the sister of Mary Camden Wetmore, who funded the Wetmore Gate ensemble. In honor of Mrs. Richards, the new gate, Beulah Road, and Beulah Lodge were all renamed Richards Gate, Richards Road, and Richards Lodge.

Walter Ratcliff's last project at Mills was construction of Mary Morse Residence Hall in 1935. Mary Morse Hall was attached to his first residence hall, Ethel Moore, constructed some ten years earlier. The two halls enclosed the landscape originally designed by Howard Gilkey for the Ethel Moore Residence Hall. This landscape was modified to accomodate the new building. Mary Morse Residence Hall is composed of two wings of student housing connected by a gracious living room and library with an asymmetrical entrance consisting of a deep-arched, inset doorway.

In 1938 McMinn's return-to-natives influence began to settle into the College's administrative thinking. The Buildings and Grounds Committee requested advice from McMinn and Gilkey's former University of California, Berkeley, professor John Gregg. Gregg was the resident university landscape architect and designer of the university's botanical garden. He prepared a succinct strategy for the College to

2-59

RICHARDS GATE, 1930s
Funded by alumna
Grace Camden Richards.

2-60

MARY MORSE HALL, 1950s
Walter H. Ratcliff, Jr., architect.

follow, "[a] plan for replanting which will include as basic trees the California live oak and the coast redwood, and many supplementary flowering trees. . . . As the eucalyptus tree is not a native of California and is inclined by the vigor of its growth to drive out oaks and pines and other native trees, the replanting does not include them." The committee qualified the plan with a specific exemption, "However many eucalypti will be retained for as long as it is possible to keep them alive—Eucalyptus Path, groves and clusters here and there on the campus." The revered Kapiolani Road blue gums were considered a landmark planting from the original campus days and worthy of retention.

Bringing closure to the dynamic Reinhardt period is the 1949 Alumnae House built in her honor after her death. Sited west of Lake Aliso along Leona Creek in a wooded riparian area, the sleek single-story residential building was designed by the San Francisco modernist architect Clarence Mayhew. Looking east from the living room window the spillway of Lake Aliso is visible and audible. The adjacent woodland area consists of eucalyptus and native buckeye trees interspersed with reintroduced native riparian plants. This natural area of the campus has a footpath and bridge allowing art students access to the nearby student center. The lawn, patio, and mixed plantings of rhododendrons, roses, and other plants provide an ideal setting for alumnae events, meetings, and social gatherings. It is a tranquil setting, a scene that would have pleased Aurelia Reinhardt.

The Reinhardt legacy was captured in *Fourscore and Ten Years* (Keep):

> The years between 1916 and 1943 mark a period of growth and expansion at Mills College in which the work of the college attained significance of a national and international character. These years cover the administration of Dr. Aurelia Henry Reinhardt. Her training and experience assured a clearly defined modern approach which the era demanded.

2-61

JUNIORS AT THE PUSH-BALL GAME, 1948

In a retrospective talk at a Founder's Day address at Sunnyside, on October 30, 1963, George Hedley, campus chaplain and author of *Aurelia Henry Reinhardt: Portrait of a Whole Woman*, spoke of the two men who made the campus a woodland both beautiful and instructive, Cyrus Mills and Howard McMinn. Having known McMinn, Hedley knew of his commitment to the restoration of California's natural landscape. Even before there was a full understanding of the relationship of flora to fauna or of habitat values, McMinn knew it intuitively. Hedley shared with the gathering, "Mrs. McMinn assures me that her husband considered the eucalyptus dangerous, and the acacias annoying."

Hedley concluded, "Cyrus Mills planted. Howard McMinn planted too. Their gifts remain with us, and so their giving persists from generation to generation. Thus also, less literally but not less truly, the plantings of all our Mills past endure into our present."

2-62

REINHARDT ALUMNAE HOUSE, LATE 1950s

Clarence Mayhew, architect.

Extant Historic Architecture

1 Mills Hall
2 El Campanil
3 Kapiolani Cottage
4 Sage Hall - Adminstration Building
5 Carnegie Library and Expansion
6 President's House
7 Lisser Hall
8 Ross and Larsen houses
9 Ethel Moore Hall
10 Mary Morse Hall
11 Music Building
12 Art Gallery
13 Orchard House
14 Meadow House
15 Warren Olney Hall
16 Norman Bridge Health Center
17 Alumnae Hall (Student Union)
18 Graduate House
19 Kimball House
20 Faculty Village
21 Reinhardt Alumnae House

Landscape

1 Evergreen forests
2 Open field
3 Lawn
4 Orchards
5 Other open spaces
6 Sunnyside
7 President's Meadow
8 Campus hearth
9 Lake Aliso
10 Creeks

Summits

1 Pine Top Hill – 310'
2 Prospect Hill – 210'
3 Sunnyside/Faculty Village Hill – 135'

Entries

1 Richards Gate
2 Wetmore Gate

Circulation

1 The Oval
2 Seminary Avenue
3 Post Road
4 Kapiolani Road
5 Richards Road
6 Orchard Meadow Road
7 Ming Quong Road
8 Bridges
9 Double lanes of Wetmore Road
10 Olney Road in Toyon Meadow
11 Road between Mills Hall and Sage Hall
12 Underwood Avenue
13 Pine Top Summit Road
14 Faculty Village Road (cul-de-sac)

2-63

MILLS COLLEGE, 1949
Elements of historic interest at the end of the campus' second period of historic significance.

3.
APPLYING THE PAST TO THE PRESENT
HOW THE CULTURAL LANDSCAPE INFORMS CURRENT PLANNING AND DESIGN

Understanding the campus' historic cultural landscape characteristics offers the College the opportunity to apply this knowledge to current and future planning and design. To demonstrate this principle, Mills College chose two areas of the campus that may undergo future improvements. The first area is the Oval and El Campanil. The second is the internal gateway formed by the intersection of Kapiolani and Richards roads. To demonstrate the flexibility in the use of historic values, there are two designs for each area, reflecting a range of building and parking programs. These studies exemplify how to plan and design in-campus areas of high cultural landscape value.

An academic home of simple beauty . . . wide lawns . . . broad gardens and shady pools for quiet study and friendly conversation. Pungent eucalyptus paths leading out and up to hilltop and to hall!

—AURELIA REINHARDT, 1927

Historic research and assessments for each site provided important base information that informed the conceptual plans and designs. By developing preliminary concepts and progressing through refinements, the College arrived at the architectural and landscape options contained in this chapter. Each concept retains and enhances key character-defining features while addressing future needs of the College

METHODOLOGY

Responding to the historic character of the Mills College campus, the process leading to implementation concepts involved the following steps:

Cultural Landscape Assessment

The cultural landscape assessment documents and analyzes a variety of factors, including periods of significance, overall landscape site integrity, incompatible features, and character-defining features.

Planning and Design Assessment

The planning and design assessment documents and analyzes the extant physical site, including current and desired uses and patterns of activities.

Preliminary Strategies for Treatment

Based on the information gained in the two assessments, the team developed treatment strategies defining the planning and design approach to the cultural landscape. Standard treatments, as defined in the *The Secretary of the Interior's Standards*

3-1

CELEBRATION IN THE LANDSCAPE
The design for the pond at the Music Building creates a theatrical setting that is both a celebration of the campus landscape and music.

3-2

COMPONENT LANDSCAPES

The campuswide landscape assessment of topography, vegetation, and the built environment revealed seven component landscapes.

for the Treatment of Historic Properties with Guidelines for the Treatment of Cultural Landscapes (1996), include preservation, rehabilitation, restoration, and reconstruction.

Design Studies

Based on the treatment strategies, the design studies serve as examples of how to incorporate cultural landscape values in planning for a range of program needs.

CULTURAL LANDSCAPE ASSESSMENT

This initial step in preparing a planning and design concept determines a site's cultural landscape values based on its character-defining features. The features analyzed are:

1. Spatial Organization, Land Use, and Context
   - Historic Architecture
   - Nonhistoric Architecture
2. Period(s) of Historic Significance
3. Topography
4. Historic Tree Palette
5. Circulation—Vehicular and Pedestrian
6. Significant Extant Hardscape Features
7. Natural Systems
8. Views and Vistas
9. Overall Site Integrity
10. Historic Plans, Drawings, and Photos

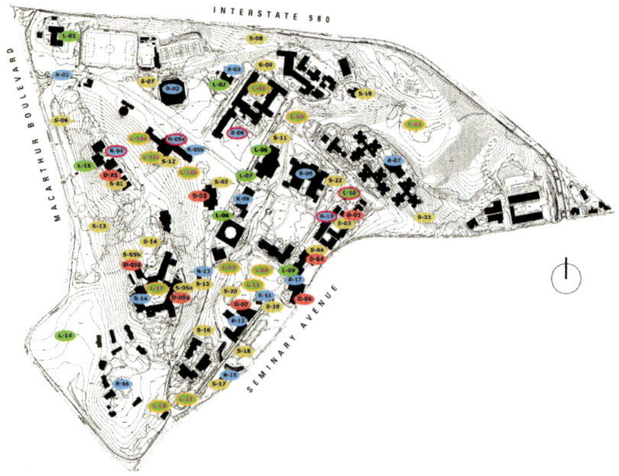

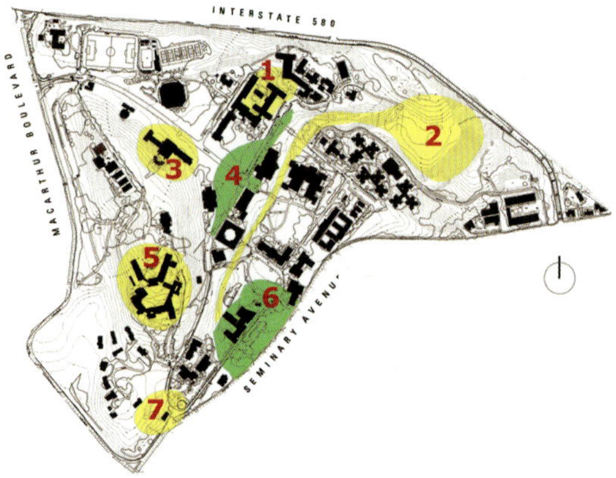

3-3

POTENTIAL IMPROVEMENTS

College staff identified numerous possibilities to consider for the design studies.

3-4

CANDIDATE SITES

The number of potential sites was narrowed to seven and then two were selected for design studies and application of cultural landscape values.

PLANNING AND DESIGN ASSESSMENT

In the planning and design assessment process, the team reviewed past planning efforts by the College, including the 2001 campus master plan. Staff discussions identified a number of critical factors that ranged from relocation and removal of facilities; pedestrian, vehicular, and service circulation patterns; and near- to long-term facility improvements. Early and frequent discussions of potential improvement projects on the campus helped determine potential sites for applying design concepts reflecting the cultural values identified by this study. This process led to the selection of the two sites discussed below. These sites have significant architecture and landscape architecture that are key to the College's image, and they involve a range of issues that the College might encounter elsewhere on the campus. For each site, the College reviewed an assessment of the character-defining features and a conceptual design framework. Further review and refinement led to the design concepts in this study.

The recognition and incorporation of historic values must take place in a broad context that recognizes and balances other, sometimes competing, needs and desires. For example in 2007, the College removed the double row of eucalyptus along Kapiolani Road in response to concerns of the age and structural integrity of the trees. Planted by the College's founder, Cyrus Mills, these eucalyptus lined a walkway and served as a defining edge to the campus before its expansion to the west. In many ways, this planting served as one of the most important character-defining features of the campus and was often promoted as an iconic symbol of Mills College.

The College's decision to remove and replant eucalyptus generated controversy about nonnative vegetation on the campus. Some members of the campus community wish to remove all eucalyptus and plant native vegetation throughout the campus, while others want to retain historic plantings of nonnatives like the eucalyptus along Kapiolani Road.

This clash of perspectives illustrates the fact that incorporating historic values into a current-day campus can be challenging. Although the College supports the removal of a majority of the eucalyptus on the campus, the restoration of the creeks, and the planting of native vegetation, it also recognizes the historic values created by its built landscape and the power of identity this imparts to the College community.

3-5

KAPIOLANI ROAD, 2006
Bryant Path in its maturity.

3-6

KAPIOLANI ROAD, 2007
Shortly before the completion of the tree removal.

STRATEGIES FOR TREATMENT

The goal of any treatment program is the long-term preservation of a landscape's historic features, qualities, and materials. Generally, the amount of physical intervention in a landscape increases from preservation to reconstruction. Preservation attempts to maintain a landscape in its existing state. Restoration often involves removing later additions and reconstructing missing features to depict a landscape at a particular time. Reconstruction replicates a nonsurviving landscape through new construction. The assessment of the two sites applied the following National Park Service definitions for the treatment of cultural landscapes:

Preservation

Preservation refers to applying measures necessary to sustain the existing form, integrity, and materials of an historic property. Work, including preliminary measures to protect and stabilize the property, generally focuses on the ongoing maintenance and repair of historic materials and features rather than extensive replacement and new construction. New exterior additions are not within the scope of this treatment; however, the limited and sensitive upgrading of mechanical, electrical, and plumbing systems and other code-required work to make properties functional are appropriate within a preservation project.

Rehabilitation

Rehabilitation refers to making a compatible use for a property through repair, alterations, and additions while preserving those portions or features that convey its historic, cultural, or architectural values.

3-7

RESEARCH, 2006
In addition to academic resources, the research included reviewing numerous plans kept in the College Facilities Department.

3-8

WORK SESSIONS, 2006
Frequent meetings of staff and consultants kept the dialogue fluent and current.

Restoration

Restoration refers to accurately depicting the form, features, and character of a property as it appeared at a particular period of time by removal of features from other periods and reconstruction of missing features from the restoration period. The limited and sensitive upgrading of mechanical, electrical, and plumbing systems and other code-required work to make properties functional are appropriate within a restoration project.

Reconstruction

Reconstruction refers to depicting, by new construction, the form, features, and detailing of a nonsurviving site, landscape, building, structure, or object to replicate its appearance at a specific period of time and in its historic location.

DESIGN PROCESS

The two selected sites—The Oval–El Campanil and Richards and Kapiolani roads—allowed staff to apply the findings of this cultural landscape assessment for both current and future design.

Several design iterations and refinements were undertaken for each of the sites resulting in the design concepts presented. The ultimate goal of this work is to guide campus improvements that build upon and strengthen the identity and mission of Mills College.

3-9

DESIGN STUDIES

Of the seven sites, two were chosen for their range of characteristics, challenges, and importance to the image of Mills College.

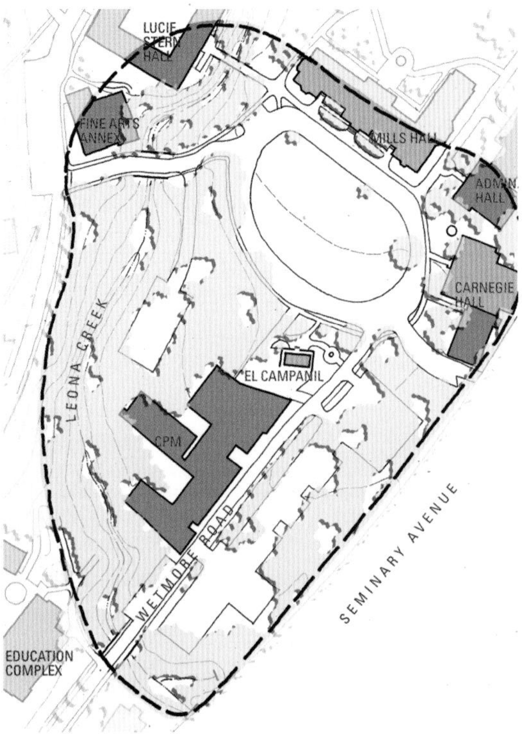

3-10

THE OVAL–EL CAMPANIL STUDY AREA

The study area includes Mills Hall, the Oval, El Campanil, Leona Creek, and a portion of Wetmore Road.

THE OVAL–EL CAMPANIL

CULTURAL LANDSCAPE ASSESSMENT

The following assesses the character-defining features of the Oval–El Campanil design study area.

1. Spatial Organization, Land Use, and Context

The Oval–El Campanil area centers on the iconic Oval, the Victorian-era landscape frontispiece to Mills Hall designed by Cyrus Mills in the late 19th-century picturesque landscape style. The study area includes the buildings and landscapes that ring the Oval and those within the western boundary of Leona Creek and Seminary Avenue and south to the confluence of Leona and Chimes Creeks. This area holds the earliest historic, character-defining features of the core campus. Encircling the Oval are some of the architect Julia Morgan's finest works: El Campanil (1904) in the Mission revival style, and the Margaret Carnegie Library (1906) in the beaux-arts style. The asymmetrical siting of the Oval buildings illustrates the deference to a pre-existing natural landscape element.

Beyond the Oval, the land is bounded by Leona and Chimes Creeks. The original, Victorian-era carriage entry to the campus was immediately off the Oval over a stylized wooden bridge from Seminary Avenue. Then just a dirt road, Seminary Avenue eventually led to a one-mile, eucalyptus-lined carriage ride that ended at the San Leandro railroad station.

Just adjacent to the south end of the Oval–El Campanil area are features from the Maybeck-influenced campus master plan of the late 1910s, later refined by campus architect Walter Ratcliff, Jr., in the mid-1920s. The grand southern campus entry at Wetmore Gate was a significant classical feature that elevated and expanded the campus grounds. The Wetmore ensemble consists of a classically designed stone and iron gate, a contiguous lodge house, and a large reflecting pool. As a narrow lane, Wetmore Road provides a slow, revelatory ascent from the gate to the Oval and Mills Hall.

The Oval–El Campanil area contains the most significant expanse of Leona Creek. A series of some 10 early-century bridges once spanned the riparian creek corridor. The bridges accommodated both pedestrians and vehicles, connecting the campus core to the later-developed west campus. All historic bridges are gone; some replaced with new, more modern pathways.

Historic Architecture within and adjacent to the Oval–El Campanil:

- Mills Hall, 1871, Second Empire Victorian style, Bugbee & Son, architects
- Sage Hall, 1882, , modest Second Empire Victorian style, architect unknown (building later subsumed by Administration Hall)
- Kimball House, 1886, architect unknown, moved to current site in 1921
- Victorian cottages, Larsen and Ross houses, circa 1871, architects unknown
- El Campanil, 1904, Mission revival style, Julia Morgan, architect
- Margaret Carnegie Library, 1906, classical/beaux-arts style, Julia Morgan, architect
- Wetmore Gate and Reflecting Pool, 1926, classical/beaux-arts style, Walter Ratcliff, Jr., architect, Mary Camden Wetmore, donor
- Administration Hall, 1930, Spanish eclectic style, Walter Ratcliff, Jr., (building subsumed Sage Hall)

Nonhistoric Architecture within and adjacent to the Oval–El Campanil:

- Education Complex, dedicated 2000, neo-Spanish colonial revival style
- Kirsch Building, 2004, neo-Spanish colonial revival style
- Chemistry, Physics, Math and Computer Science, dedicated 1971, contemporary style
- Southern Annex to Carnegie Library, 1954, contemporary style

2. Periods of Historic Significance

1868–1916

Dr. Cyrus and Susan Mills

English country estate/picturesque era

1916–1949

Dr. Aurelia Reinhardt, President

Walter Ratcliff, Jr., Campus Architect

Spanish colonial revival/eclectic era

3. Topography

The Oval and the site for Mills Hall were once an open meadow flanked by riparian oaks along Leona and Chimes creeks. The land to the south and the distant train station inspired a south-facing orientation for Mills Hall. The topography is a consistent, moderately sloped ascent of approximately 30 feet from Wetmore Gate to Mills Hall.

4. Historic Tree Palette

The campus landscape is characterized by a generally monochromatic tall evergreen tree canopy. Eucalyptus predominates; it is found in almost every corner of the campus. The Oval–El Campanil area holds some of the first plantings on the campus. Eucalyptus is also present along most of the length of Leona Creek.

Remnant groves of the eucalyptus and pines planted by Cyrus Mills have exceeded their original Victorian-era boundaries and intent. Early groves were located south of El Campanil, on much of the original southern campus, behind the Natural Sciences Building in the areas north and south of Lake Aliso, and on the lands flanking the California Railway Streetcar K-Leona station, where interstate 580 and MacArthur Boulevard are today.

Between Sage Hall and Mills Hall was an expressive formal parterre floral garden that according to secondary sources may have showcased the Mt. Vernon magnolia tree (still existing) that was given to Cyrus Mills.

At Wetmore Gate a ring of poplars stands around the circular reflecting pool, further expressing the beaux-arts style of the gate and pool ensemble. Landmark mature trees on the north and west sides of Mills Hall are noted in the 1996 Beatty *Vegetation Management Plan*. These mature specimens are the exotic *Araucaria* species, a New Zealand favorite planted throughout California in the late 19th century. The front fenestration of Mills Hall is obscured by recent, irregularly placed and rigidly maintained nonhistoric plant additions. Missing are the signature rose climbers from the period of significance that were once planted on the Rose Porch.

An historic tree palette should be further developed for future interpretation of the period within historically sensitive areas.

5. Circulation—Vehicular and Pedestrian

Historic Wetmore Road still provides a narrow drive connecting the 1926 formal entry at Wetmore Gate to Mills Hall and the Oval. One lane remains from a north-south two-lane boulevard that held a generous median. Born from Maybeck's north/south axis, later refined in Ratcliff's mid-1920s plan for the campus, this boulevard ended at El Campanil at the north end and the Wetmore Pool at the south. The western segment of the boulevard remained in place through the 1970s. The surviving eastern portion retains its original alignment.

Asphalt pedestrian walks parallel most of the west bank of Leona Creek. Other pedestrian walks are specific to buildings and destinations and are not designed as campuswide looped walks. The traffic noise of Seminary Avenue prominently affects this area, especially at Wetmore Gate and the Education Complex/Children's School. Exposed surface parking lots line the east side of Wetmore Road and create visual

3-11

WETMORE GATE, 1930s
Wetmore Gate viewed from the reflecting pond.

3-12

WETMORE ROAD, LATE 1920s
This double-lane road connected the Oval through a parklike setting to the new south entry, Wetmore Gate.

3-13

WETMORE ROAD, 2007
Today, Wetmore Road is reduced to one lane and lacks its past parklike character.

clutter in this once-serene parklike setting. Identification of buildings, pedestrian and vehicular routes, and natural elements of the campus is not sufficiently visible. Comprehensive directional and building signage is needed campuswide.

6. Significant Existing Hardscape Features

Wetmore Gate and the circular reflecting pool (approximately 25 feet in diameter) were designed in the beaux-arts tradition. Bridges over Leona and Chimes creeks are nonhistoric replacements. The small fountain on the front lawn of Carnegie Hall dates back to the first period of significance but has lost much of its detail.

7. Natural Systems

Leona Creek and Chimes Creek are open, riparian environments with excessive nonnative plants and seasonal creek scouring. Chimes Creek is buried in a culvert east of the Natural Sciences Building complex, resurfacing where it joins Leona Creek, north of Kimball House. This confluence is obscured by vegetation and nonhistoric road infrastructure. The open meadows and oak woodland that once defined the Victorian-era landscape has been overtaken by unmanaged exotic tree proliferation, predominantly eucalyptus. Consequently, the original land characteristics that attracted Cyrus and Susan Mills to the property have been significantly compromised.

8. Views and Vistas

Distant picturesque views are obscured by the mature tree canopy. This outer circle of trees defines the campus' open space, harkening back to the original open meadows of the core campus.

Limited views within the riparian corridor of Leona Creek are more shaded than would occur naturally. Upper Chimes Creek is oppressively dark in places.

The open chain-link fence that runs from Wetmore Gate to the north does not screen views of traffic and the adjacent commercial neighborhood. The College should study view mitigation, sound attenuation, and security to return to a quiet aesthetic ambience in this historic area.

Views of significant historic architecture are obscured and diminished. For example, the size and proximity of the 1971 Chemistry, Physics, Math and Computer Science building complex (CPM) disturbs views of El Campanil from both Wetmore Road and the Oval.

9. Overall Site Integrity

Incompatible architecture and inappropriate site planning compromises the Oval-El Campanil area. The closed Wetmore Gate imparts a ghost-town effect to the area.

Vistas and views are major character-defining features of picturesque era landscapes. The placement of historic buildings relied on this land-and-view ideal. Today, the closely planted mature tree canopy has grown beyond the original design intent. To reclaim the original character of the grounds, the College should consider select removal, in particular removal of exotic species, to open views. Also, this effort would address public safety and the College's desire for a responsible balance of native and nonnative species.

Transportation and traffic have detracted from the campus. The College should consider future campus projects that better preserve the atmosphere of educational retreat intended by Cyrus and Susan Mills.

10. Historic Plans, Drawings, and Photos

- *Contour Map of Mills College Grounds, Oakland, Alameda County, California,* November, 1916. E. C. Prather, Civil Engineer.

3-14

CARNEGIE FOUNTAIN, CIRCA 1893

A mature garden parterre surrounds the fountain.

3-15

CARNEGIE FOUNTAIN, 2007

Over 100 years later, the fountain requires restoration including reestablishment of the garden.

- *Contour Map of Mills College Grounds, Oakland, Alameda County, California*, December, 1928. W. H. Ratcliff, Jr., Architect. C. L. Huggins, and Howard Gilkey. Revised January, 1930.
- Aerials and other select photos of the area are available.

PLANNING AND DESIGN ASSESSMENT

The Oval and El Campanil form the historic core of the Mills College campus. Early photographs of Mills Hall fronted by the oval of lawn captured the essence of the campus in the late 1800s. The Oval is a large elliptical plane of lawn measuring approximately 260 by 160 feet. This welcoming carpet of green without trees and shrubs affords unencumbered views of Mills Hall. Julia Morgan's Carnegie Library lies to the east in close proximity to Seminary Avenue, which forms the noisy eastern edge of the campus. El Campanil lies across the lawn from Mills Hall. Along with the axis of Mills Hall, those of El Campanil and the Carnegie Library intersect in the Oval.

When first constructed, El Campanil was part of a larger landscape setting. However, over the decades that followed, the encroaching built environment reduced the visual foundation provided by the surrounding landscape. El Campanil today stands out as a singular object in a hardened environment. In contrast, the native oaks and introduced vegetation along Leona Creek frame the southern edge of this area.

Reducing the quality of the pedestrian experience, cars freely navigate and park around the Oval, circulating east over Leona Creek or south toward the Wetmore Gate. Wetmore Road, once part of a pair of roads leading through a parklike setting from Wetmore Gate to the Oval, is currently purely functional and detracts from the historic landscape of the campus.

The campus master plan of 2001 recommends the intensification of uses in the vicinity of the Oval, and the recently completed creek study calls for the protection and enhancement of Leona Creek. Many have expressed the desire to return this

3-16

EL CAMPANIL, CIRCA 1910
Shortly after its completion, El Campanil was integral to the surrounding landscape.

3-17

EL CAMPANIL, 1925
The landscape remained a powerful partner while the automobile began its intrusion on campus.

3-18

EL CAMPANIL, CIRCA 1970
By the 1970s, the surrounding landscape was greatly diminished and the automobile has become a dominant element on campus.

3-19

EL CAMPANIL, 2006
Today, El Campanil is setting in a hard urban plaza lacking mature plantings—the antithesis of the original design intent.

area to a pedestrian environment, relegating the automobile to other areas of the campus. As part of the planning process, the College identified several buildings that are likely to be removed in the long term, providing the opportunity to locate and design new buildings in a manner respectful of the College's historic values.

The design concepts for this study area explore two possible programs. The first addresses the introduction of new buildings that can house a range of academic and administrative uses. The second explores new facilities with increased parking east of Wetmore Road, as recommended in the campus master plan. Although both schemes are valid, the concept that includes parking raises concerns about campus circulation and security in the specific area of the campus that the College desires to become pedestrian. This issue needs to be explored as part of a broader, campuswide planning study.

Because they are outdated and lacking historic value, these buildings will eventually be relocated or demolished:

1. Lucie Stern Hall
2. Fine Arts Annex
3. Sage Hall (Administration Building)
4. The 1950s addition to Carnegie Hall
5. Chemistry, Physics, Math, and Computer Science (CPM)

PRELIMINARY STRATEGIES FOR LANDSCAPE TREATMENT

The primary landscape treatment for the Oval–El Campanil area is rehabilitation, typically the treatment applied to compromised historic landscapes. The landscape treatment strategy also identified selected areas for restoration and reconstruction.

Mills Hall Facade—Restoration and Reconstruction

- Reintroduce the Rose Porch with heirloom rose species at the primary entry to Mills Hall; use a support structure separate from the building and manage rose growth.
- Reconstruct the missing cupola atop Mills Hall.
- Restore the historic paint color palette to Mills Hall; use forensic testing to determine appropriate period colors.

The Oval and El Campanil—Restoration, Rehabilitation, and Reconstruction

- Reintroduce the historic path across the Oval from Mills Hall to El Campanil.
- Reintroduce or reinterpret the original vehicular entry adjacent to Carnegie Hall that led from Seminary Avenue over a bridge to the Oval.
- Remove the southern addition to Carnegie Hall and restore Carnegie Hall's south elevation.
- Remove Sage Hall (Administration Building) to reintroduce the parterre garden in and around the original 1882 Sage Hall. Represent the original Sage Hall with an open pavilion structure.

3-20

THE OVAL, 2006
Automobile parking visually impairs the once bucolic setting of Mills Hall.

3-21

MILLS HALL, 2007
Preliminary strategies for landscape treatment include reconstruction of the cupola and reintroduction of the Rose Porch.

- Establish a setback around the entire Oval for new buildings, consistent with the existing setback established by Carnegie Hall.
- Establish a pedestrian environment for the Oval; remove parking and reclaim the area for an enhanced landscape setting.
- Establish a setback of 100 feet around El Campanil to reestablish the character of the original landscape setting, and return hardscape elements to their original configurations.
- Realign the road and rebuild the bridge entering the west side from Kapiolani Road to the Oval.

Wetmore Road—Rehabilitation

- Reintroduce the informal parklike setting of Wetmore Road with informal planting and a varied setback for new buildings.
- Create a shared landscape to visually connect Wetmore Gate to the Oval.
- Relocate the historic Ross and Larsen houses in proximity to the Kimball House, creating a Victorian-era ensemble.

Leona Creek Interface—Restoration

- Restore the Leona Creek riparian and upland corridor according to the 2007 creek study.

Chimes Creek—Reconstruction

- Uncover and reconstruct Chimes Creek, currently buried.
- Reconstruct bridge.

Plantings

- Selectively remove plantings that lack historic value.
- Use native plant materials.

New Design

- Explore site-sensitive building development opportunities in the vicinity of the Chemistry, Physics, Math, and Computer Science building complex.
- Explore site-sensitive building development opportunities in the area east of Wetmore Road.
- Explore period-compatible design for the bridge crossing at the confluence of Leona and Chimes Creeks.
- Establish native plants, removing the majority of nonnative plant material.
- Explore a sound attenuation wall along Seminary Avenue.

3-22

HISTORIC ENTRY, CIRCA 1883

Native vegetation framed the historic entry from Seminary Avenue with an expansive view of Mills Hall beyond.

3-23

HISTORIC ENTRY LOST, 2007

One strategy is to reintroduce or reinterpret the original entry that today is lost to parking and a building addition.

Character-defining features of the Oval-El Campanil study area:

1. The Oval
2. Circular drive
3. Original entry
4. Asymmetrical siting of buildings facing the Oval
5. 360 degree view of El Campanil
6. Leona Creek
7. Parkway entry from Wetmore Gate

Design objectives for the Oval-El Campanil study area:

1. Pedestrianize the Oval.
2. Reestablish the oak grove around El Campanil.
3. Reestablish the original entry to campus.
4. Reintroduce the parterre garden.
5. Replace bridge per the creek study.
6. Redesign bridge to emphasize the crossing of Leona Creek.
7. Reintroduce an informal parklike setting along Wetmore Road with informal tree planting and varied setbacks of new buildings.
8. Restore Leona Creek.
9. Expose Chimes Creek, currently buried.
10. Explore new building opportunities.

3-24

PRIMARY CHARACTER-DEFINING FEATURES

Based on the landscape assessment, the study area has seven primary character-defining features.

3-25

DESIGN FRAMEWORK

Analysis of the character-defining features, current and future campus functions, and future opportunities yielded ten design objectives to be incorporated into the design studies.

DESIGN CONCEPT

The following pages illustrate possible design concepts for the Oval–El Campanil area based on the cultural landscape assessment, the planning and design assessment, and the strategies for treatment. The concepts reflect two programs. The first includes only buildings for academic and administrative uses. The second includes buildings for academic and administrative uses and a new parking structure. The two schemes serve to illustrate the flexibility of incorporating cultural historic values in current campus planning and design.

3-26

VIEW OF EL CAMPANIL FROM THE OVAL

Both design studies promote the reestablishment of a strong landscape foundation for El Campanil while introducing new facilities that are sympathetic to this historic setting.

3-27

THE OVAL–EL CAMPANIL— NEW FACILITIES AND PARKING

This scheme allows continued but more controlled use of the area by automobiles.

3-28

THE OVAL–EL CAMPANIL— NEW FACILITIES

This scheme reclaims the area for pedestrian use thereby greatly expanding land available for landscape improvements.

3-29

REINTRODUCED ENTRY

Both schemes propose the reintroduction of the original entry as viewed from Seminary Avenue just south of Carnegie Hall. This includes a bridge and the restoration of a portion of Chimes Creek, currently buried.

3-30

GARDEN PARTERRE AND SAGE HALL

As viewed from the eastern side of Mills Hall, the design suggests the restoration of Carnegie Fountain, the reestablishment of the parterre garden, and construction of a "ghost structure" that would follow the massing and style of the original Sage Hall.

Shared elements in the two schemes:

1. Reestablish the Rose Porch at Mills Hall with a metal frame support structure to separate the vines from the building.
2. Remove parking along the Oval, resulting in an increase in landscaped area on its outside edge.
3. Allow 24 feet for road width next to Mills Hall to accommodate both vehicular circulation and temporary parking for deliveries.
4. Expand parking for the handicapped and introduce two visitor parking spaces at the existing parking area northwest of the Oval.
5. Reestablish a path crossing the Oval connecting to El Campanil and its adjacent buildings.
6. Improve Leona Creek according to the recent creek study and reestablish paths along the creek.
7. Replace and realign the bridge crossing Leona Creek as recommended in the recent creek study.
8. Replace the bridge as it crosses Wetmore Road at the confluence of Leona and Chimes creeks with a period-compatible design.
9. Reintroduce a symbolic and possibly functional entry from Seminary Avenue at the original entry to the Mills campus to include a bridge and the restoration of Chimes Creek in this area.
10. Reestablish the informal parklike planting around El Campanil and along Wetmore Road.
11. Relocate two historic buildings, Larsen and Ross houses, from the area north of the science complex to an area north of Leona Creek at Seminary and restore Chimes Creek and a pedestrian bridge to access the relocated buildings.
12. Reestablish the parterre garden and construct a "ghost structure" recalling Sage Hall between Mills Hall and Carnegie Hall.
13. Establish a sound wall along Seminary Avenue.
14. Introduce two new buildings west and south of El Campanil.

Elements unique to the
New Facilities and Parking scheme:

1. Two new buildings west and south of El Campanil—approximately 40,000 gross square feet (gsf) at two stories, 60,000 gsf at three stories
2. A new building at the northeast side of Wetmore Road—approximately 17,000 gsf at two stories
3. A new parking structure with internal speed ramps along the eastern edge of Wetmore—approximately 200 cars
4. A new building to wrap the northern edge of the new parking structure —approximately 8,000 to 12,000 gsf at two and three levels, respectively

Total Capacities:

- Building capacity at two stories —approximately 65,000 to 69,000 gsf
- Parking capacity at three levels —approximately 200 cars

Elements unique to the
New Facilities scheme:

1. Two new buildings west and south of El Campanil—approximately 40,000 gsf at two stories, 60,000 gsf at three stories
2. Two new buildings along the east side of Wetmore Road—approximately 44,000 to 50,000 gsf at two to three stories with the portion of the building closest to El Campanil kept at two stories to ensure visibility of the historic structure

Total Capacities:

- Building capacity at two to three stories —approximately 88,000 to 110,000 gsf
- Parking capacity —modest allocation for visitors and handicapped

3-31

SHARED ELEMENTS

Many of the elements are common to both design schemes.

3-32

NEW FACILITIES AND PARKING UNIQUE ELEMENTS

Several elements are unique to this design scheme.

3-33

NEW FACILITIES UNIQUE ELEMENTS

Several elements are unique to this design scheme.

3-34

RICHARDS AND KAPIOLANI ROADS STUDY AREA
The study area includes the intersection of Richards and Kapiolani roads, the F. W. Olin Library–Lisser Hall courtyard with Leona Creek beyond, and the site currently occupied by the Cowell Building.

RICHARDS AND KAPIOLANI ROADS

CULTURAL LANDSCAPE ASSESSMENT

The following assesses the character-defining features of the Richards and Kapiolani roads design study area.

1. Spatial Organization, Land Use, and Context

The Richards and Kapiolani roads area includes Kapiolani Road from the Old Post Road to Lucie Stern Hall; Richards Road from the Music Building to its intersect with Kapiolani Road; and the adjacent meadows, slopes, and mature landscapes. It also encompasses the open plaza between the historic Lisser Hall and F. W. Olin Library, as well as their interface with Leona Creek.

Richards Road (circa mid-1920s), formerly Beulah Road, is reminiscent of a European boulevard. It was included in Allan B. Jacobs' book *Great Streets* as one of the great avenues in the world. London plane trees (sycamore), *Platanus acerifolia*, are planted in triangulated rows on each side of Richards Road in generous twenty-foot parkways that flank informal walking paths . The method of early pruning doubled the tree canopy, creating a wide arc of foliage. The expansive meadows along Richards Road enhance the power of this planting.

This grand, ordered streetscape is the iconic expression of the campus' western expansion in the 1920s. Its design is a departure from the original campus aesthetic to the east, which is predominantly picturesque. Well past a century of maturity, the old eucalyptus path from the first period of significance marked the western edge of the original campus. Perpendicular to Richards Road, eucalyptus towered above all natural and built elements. This double soldier-row planting of blue gum eucalyptus, wildly popular during California's frontier days, marked the remains of Bryant Path planted by Cyrus Mills in the late 19th century.

The plantings of Richards and Kapiloani roads symbolized the meeting of the new and the old campus. In ancient Persia, deciduous trees represented life cycles and the female value while evergreen trees represented the eternal and the male value. It is not known if Bernard Maybeck or Howard McMinn had this mythology in mind when they chose London plane trees for Richards Road, but the contrast with the eucalyptus of Kapiolani Road was striking and it marked the historic epicenter of the campus.

The architecture of Arthur Brown, Jr., of Bakewell & Brown, and Walter H. Ratcliff, Jr., illustrate their sensitivity to the ground plane, slopes, and other characteristics of the topography. In Ratcliff's siting, the Music Building runs parallel to and is nestled in the slope of the hillside with the Greek Theatre set on a cross axis to the longitudinal front elevation of the building. Howard Gilkey designed informal yet classical details for the landscape along the front of the Music Building and outdoor theatre. Howard McMinn reintroduced the oak woodland along the slopes above Ratcliff's Music Building. Future campus improvements should be sensitive to the accomplishments and visions of these design masters.

Historic Architecture

- Lisser Hall, 1901, Greek Revival style, associated with Willis Polk, architect; remodel, 1927, Spanish colonial revival style, Walter Ratcliff, Jr., architect
- Meadow Hall, 1921, arts and crafts style and Spanish colonial revival style, Bakewell & Brown, architects
- Music Building and Greek Theatre, 1928, Walter Ratcliff, Jr., architect and Howard Gilkey, landscape architect
- Richards Road tree planting, early 1920s, Bernard Maybeck and Howard McMinn

Nonhistoric Architecture

- Cowell Building, 1966
- F. W. Olin Library, 1990
- Lucie Stern Hall, 1965
- Graduate School of Business, 2009

3-35

EUCALYPTUS PATH, 2006
This eucalyptus path once towered over the campus landscape.

3-36

EUCALYPTUS PATH, 2007
The trees' uncontrolled buttress roots greatly diminished the usability of the path.

3-37

CLOSE-UP OF EUCALYPTUS BARK
The bark of the blue gum eucalyptus is distinctive in texture and color.

2. Periods of Historic Significance

The Richards and Kapiolani roads area has two periods of significance:

1868–1916

Cyrus and Susan Mills

English country estate/picturesque era

1916–1949

Aurelia Henry Reinhardt, President

Walter Ratcliff, Jr., Campus Architect

Spanish colonial revival /eclectic era

Once known as Bryant Path, the eucalyptus walk along Kapiolani Road delineated the western boundary of the campus in the first period of significance. The plane tree allée along Richards Road represents the modern western campus in its second period of significance. Later incompatible additions occur in the F. W. Olin Library–Lisser Hall plaza area along the eastern side of Kapiolani Road as well as development of the Cowell Building.

3. Topography

The site is the flattest and most visually open area on the campus. The exception is the dramatic slope behind the Music Building and the Cowell Building that wraps around the eastern side of the Kapiolani and Richards roads area.

4. Historic Tree Palette

The eucalyptus path predated Kapiolani Road. It was a notable cultural landscape element harkening to the first period of significance. The path provided shade and a dramatic setting for those going to the K-Leona streetcar station on the northern border of the campus. Until their removal in 2007, this swath of uniform double-row evergreen blue gum, *Eucalyptus globulus,* displayed maturity and verticality that commanded attention and admiration from both distance and nearby vantage points.

The Richards Road streetscape is one of the most valuable plantings on campus. It dates from the second historic period of significance. The allée of London plane trees, *Platanus x. acerifolia,* extends along the entire length of Richards Road. The Maybeck-

3-38

RICHARDS ROAD, 2006
Nearing 90 years of growth, the London plane trees create a monumental yet transparent frame for the campus' entry drive.

3-39

RICHARDS ROAD, 2006
Each double row of trees contains a pedestrian walk that accesses buildings located a distance from the street edge.

designed allée consists of double rows of plane trees chosen in consultation with botanist Howard McMinn.

There are several specimen trees in the area. Specimen redwoods are ranged along Leona Creek as well as grouped around Meadow Hall, the Music Building, and in front of the Cowell Building. Howard McMinn introduced native oak woodland behind the pond and the amphitheatre adjacent to the Music Building. Landmark mature trees are noted in the 1996 Beatty *Vegetation Management Plan*—for example, oak, *Quercus agrifolia*, at the Music Building and a bald cypress, *Taxodium distichum*, near Meadow Hall.

3-40

THE POND, 2007
The pond still offers the opportunity to fulfill the original design intent of Howard Gilkey.

3-41

FOUNTAIN REMNANTS, 2007
Remnants of the fountain are visible near the pond.

5. Circulation—Vehicular and Pedestrian

Richards and Kapiolani roads are the two most-traveled roads on the west campus. The sidewalk between the plane trees provides a pleasing pedestrian experience during all seasons. By contrast, the asphalt path between the eucalyptus trees became severely compromised over time by the uncontrolled buttress roots of the trees. A formal sidewalk along the east side of Kapiolani Road leads into the plaza between Lisser Hall and the F. W. Olin Library. A road winds behind the Cowell Building for service access to the Music Building and Greek Theatre.

6. Significant Existing Hardscape Features

Fountain relics adjacent to the pond and the Music Building are in disrepair. A modern sculpture, incompatible with its setting, is located in the plaza between Lisser Hall and F. W. Olin Library.

7. Natural Systems

While located at the eastern side of the study area, Leona Creek continues to be a natural setting with great potential to connect to Kapiolani Road. Howard Gilkey's pond adjacent to the Music Building takes advantage of the open views of the meadows along Richards Road.

8. Views and Vistas

Views within and around the Richards Road plane tree allée are spectacular, and the meadows adjacent to the trees expand the visual experience. The eucalyptus path was evocative of an earlier day. The opportunity to open a vista from Richards Road through the open plaza toward the old campus would have great cultural value. Visually connecting these two historic areas has the potential to unite the two eras of the campus.

9. Overall Site Integrity

Despite later architecture and compromised landscapes, this area has great historic interpretive potential. By removing inappropriate features that detract or diminish the site's character, the College can introduce elements that are compatible with earlier historic patterns.

There is historic integrity in the Richards Road landscape design with sycamore planting, parkways, and adjacent meadows, continuing its original design intent as a distinctive setting and entry experience. The oak woodland behind the Music Building planted contemporaneous to the architecture is a successful turn toward native plant restoration and reclamation of the land. Until its removal, the majestic eucalyptus path retained its place, highlighting the early history of the campus.

10. Historic Plans, Drawings, and Photos

- Refer to Section 2: Exploring the History of Campus for descriptions of Howard Gilkey's landscape plan for the Music Building, the area including the intersection of Richards and Kapiolani roads to the Music Building, the historic plane tree allée, and the eucalyptus path.
- *Contour Map of Mills College Grounds, Oakland, Alameda County, California*, December, 1928. W. H. Ratcliff, Jr., architect, C. L. Huggins, and Howard Gilkey, revised January, 1930.
- Aerials and other select photos of the area are available.

3-42

LEONA CREEK, CIRCA 1950s
Leona Creek provided shaded relief to the campus.

3-43

LEONA CREEK, 2006
Today, the creek is heavily eroded and overgrown with nonnative vegetation. A recently completed study identifies remedies to these issues.

3-44

THE INTERSECTION OF RICHARDS AND KAPIOLANI ROADS, 2007
Once important to the character of the campus, this intersection has become purely functional and is not reflective of the College's identity.

PLANNING AND DESIGN ASSESSMENT

Richards Road and Kapiolani Road, with their respective plantings of London plane and eucalyptus blue gum, once served as iconic expressions of where the old campus intersected the new. Replanting of a different eucalyptus species will reassert this historic border. Framed by Lisser Hall and the F. W. Olin Library, travelers on Richards Road still glimpse Leona Creek and Toyon Meadow beyond. Buildings along Kapiolani Road create a wall between the natural environment of the creek and the newer part of the campus punctuated by Richards Road. Past designs, such as the row of street trees along the west side of the F. W. Olin Library, are urban in character in contrast to the informal natural meanderings of the creek. The College's desire to replant a double row of eucalyptus along Kapiolani Road to replace those recently removed illustrates the College's recognition of the historic cultural importance of the design statement made by Cyrus Mills. Discussion with staff of the Presidio Trust identified *Eucalyptus saligna*, Sydney blue gum, as an appropriate in-kind replacement species because of its color and growth characteristics.

The area surrounding Richards Road is gaining academic importance as evidenced by the renovation of the Music Building and the planned construction of a new building to house the Graduate School of Business. The latter is sited in the open meadow at the northwest corner of Richards and Kapiolani roads. These facilities, along with

existing student housing, emphasize the need for better pedestrian connections between the historic core and the west campus. The removal of parking along Kapiolani Road in the vicinity of the Olin Library will allow the College to narrow the width of the road and increase the pedestrian character in the area.

The recent creek study emphasizes the need to remove and replace nonnative vegetation with native plants appropriate to the creek environment. In concert with the creek improvements, the College intends to add and refurbish trails that bound each side of the creek. The College also hopes to uncover the long-buried Beulah Brook that traversed the area now occupied by the Cowell Building. Lisser Hall, identified by the College for future renovation, will require access for the handicapped.

Because it is outdated and lacks historic value, the Cowell Building will eventually be demolished. The College plans to explore potential redevelopment in this area.

3-45

F. W. OLIN LIBRARY–LISSER HALL COURTYARD, 2007
Viewed from the entry to Lisser Hall, the courtyard is characteristically urban, lacking a physical and aesthetic connection to the natural characteristics of nearby Leona Creek.

3-46

VISUAL CONNECTION NEEDED, 2007
Visual connections to Leona Creek and the native vegetation are blocked by over-sized planters and urban tree plantings.

3-47

SIMPLE PAVING, 2007
The outdoor lobby to Lisser Hall offers an example of the power and simplicity of concrete as a paving material.

PRELIMINARY STRATEGIES FOR LANDSCAPE TREATMENT

The overall primary landscape treatment for the Richards and Kapiolani roads area is reconstruction and rehabilitation.

Kapiolani Road—Reconstruction

- Replace in kind the historic eucalyptus path in its original alignment.
- Select as replacement species a eucalyptus genus exhibiting similar characteristics to *Eucalyptus globulus*: vertical, tall, evergreen, smooth light-colored bark, similar sickle leaf shape.
- Use an approximate spacing of 12 by 12 feet, similar to the original planting but accommodating to future tree growth and pedestrian access.
- Use natural materials for path, such as decomposed granite.
- Remove parking along a portion of Kapiloani Road, narrowing the roadway.
- Extend the creek planting through the Lisser-Olin courtyard and along the eastern edge of Kapiolani Road.

Richards Road—Preservation and Reconstruction

- Minimize changes to the landscape of Richards Road; maintain the existing rows of London plane trees and adjacent lawn.

- Preserve the open character of the meadows along Richards Road with a building setback and planting; respect the setback/build-to line established by the Music Building and the Orchard Meadow Residence Hall.
- Reestablish the Howard Gilkey design for the pond adjacent to the Music Building.

Intersection of Kapiolani and Richards Roads—Rehabilitation

- Use design cues from the 1922 Maybeck-McMinn landscape plan.
- Reestablish the historic sense of arrival and character at the intersection of Richards and Kapoiolani roads.
- Introduce directional signage in the area.
- Retain open space at the two western corners of the Richards and Kapiolani roads intersection.

Beulah Brook—Reconstruction

- Uncover Beulah Brook to become a tributary to Leona Creek.

Plantings

- Selectively remove plantings that lack historic value.
- Use native plant materials.

New Design

- Reconcile the disparity between the facades and geometry of historic Lisser Hall and F. W. Olin Library.
- Introduce interpretive elements to express character-defining features from both periods of significance, such as Spanish eclectic/classical facade and columns of Lisser Hall and the natural character of Leona Creek and its immediate environs.
- Use design, bridge reconstruction, and selective thinning of vegetation to visually connect Richards Road to Leona Creek, Toyon Meadow, and beyond.
- Explore potential building development in the location of the Cowell Building.

Character-defining features of the Richards and Kapiolani roads study area:

1. Double row of deciduous trees
2. Generous setback to building facades
3. Simple treatment of lawn-meadow in setback
4. Open lawn-meadow at intersection of Richards and Kapiloani roads
5. Pond
6. Double row of evergreen (eucalyptus) trees
7. Functional and ceremonial node
8. Views to Leona Creek and beyond
9. Natural vegetation
10. Leona Creek

Design objectives for the Richards and Kapiolani roads study area:

1. Touch lightly.
2. Maintain setback for future buildings.
3. Maintain and enhance open green at corners.
4. Reestablish the double row of evergreen (eucalyptus) trees.
5. Complete the pond design.
6. Replace the bridge per the creek study.
7. Reestablish the ceremonial node.
8. Evaluate the courtyard's circulation and design.
9. Extend natural planting from Leona Creek to southern edge of Kapiolani Road.
10. Restore Leona Creek.
11. Explore new building opportunities.
12. Explore expansion opportunity for F. W. Olin Library.

3-48

PRIMARY CHARACTER-DEFINING FEATURES

The landscape assessment revealed seven primary character-defining features in the study area.

3-49

DESIGN FRAMEWORK

Analysis of the character-defining features, current and future campus functions, and future opportunities yielded twelve design objectives to be incorporated into the design studies.

DESIGN CONCEPT

The following pages illustrate possible design concepts for the Kapiolani and Richards roads area based on the cultural landscape assessment, the planning and design assessment, and the strategies for treatment. The concepts reflect two programs. The first includes only buildings for academic and administrative uses. The second includes buildings for academic and administrative uses and a new parking structure. The two schemes illustrate the flexibility of incorporating cultural historic values in current campus planning and design.

3-50

NEW ENTRY AT RICHARDS AND KAPIOLANI ROADS

Shared between the two design schemes is the redesign of this internal campus entry. The proposed design includes lighted columns in character with other past and present campus entries, a new shelter for shuttle users inspired by the shape of eucalyptus leaves, and the interweaving of native vegetation to tie this area to Leona Creek.

3-51

RICHARDS AND KAPIOLANI ROADS—NEW FACILITIES AND PARKING

This scheme includes both new facilities and a small parking structure wrapped by two buildings.

3-52

RICHARDS AND KAPIOLANI ROADS—NEW FACILITIES

This scheme focuses on new facilities and further limiting on-street parking. As with the prior scheme, this design includes the reestablishment of the eucalyptus path, the redesign of the courtyard between the F. W. Olin Library and Lisser Hall, and the completion of the design for the pond next to the Music Building.

Shared elements in the two schemes:

1. Narrow the north portion of Kapiloani Road and eliminate street parking.
2. Narrow the intersection of Kapiolani and Richards roads.
3. Replace original planting using *Eucalyptus saligna* along Kapiolani Road.
4. Raise all pedestrian crossings to the level of the adjacent pedestrian paving.
5. Expand the size of the existing pond east of the Music Building to better reflect the intent of the designer, Howard Gilkey; include a Temple of Music in deference to the Temple to the Radio originally envisioned by Gilkey.
6. Introduce a low fence at the intersection of Richards and Kapiolani roads to direct views and pedestrian traffic into the historic core of the campus.
7. To accommodate shuttle-bus users, introduce seating that is partially covered and faces Richards Road.
8. Redesign the open space between Lisser Hall and the F. W. Olin Library to accommodate pedestrian movement patterns and to diminish the incompatible visual image of the F. W. Olin Library façade.
9. Rebuild the bridges.
10. Accommodate the possible. northern expansion of the F. W. Olin Library —approximately 10,000 gsf at two stories.
11. Uncover Beulah Brook.
12. Develop new buildings in the area southwest of Richards and Kapiolani roads.

Elements unique to the
New Facilities and Parking scheme:

1. Buildings wrapping parking structure on north and east sides —approximately 26,000 gsf at two stories
2. Parking structure with internal speed ramps
3. Service road to the Music Building realigned into the hillside

Total Capacities:
(including expansion of the F. W. Olin Library):

- Building capacity at two stories —approximately 36,000 gsf
- Parking capacity at three levels —approximately 125–135 cars

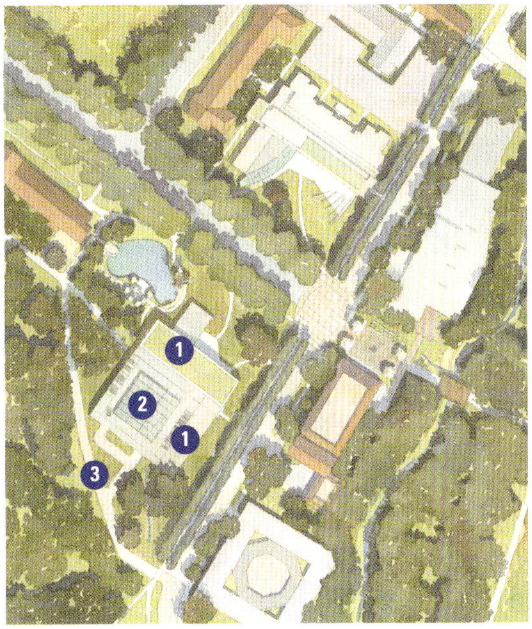

Elements unique to the
New Facilities scheme are two buildings with a shared courtyard:

1. The western building is sited along the contours of the hillside; the building has a large courtyard built into the hillside—approximately 26,000 gsf at two and a half stories.
2. The eastern building frames the expanded pond and parallels Richards Road; its wing extends toward Richards Road and aligns with the built line of the Music Building—approximately 20,000 gsf at two stories.

Total Capacities
(including expansion of the F. W. Olin Library):

- Building —approximately 56,000 gsf
- Parking —existing along Richards Road and north of the Graduate School of Business

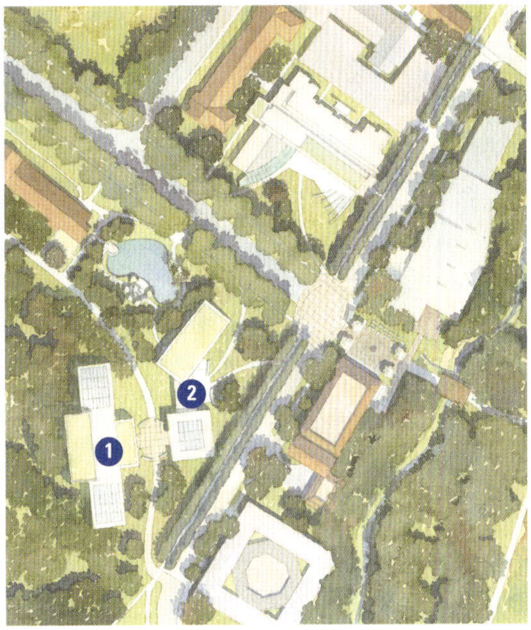

3-53

SHARED ELEMENTS
Many of the elements are common to both design schemes.

3-54

NEW FACILITIES AND PARKING UNIQUE ELEMENTS
Several elements are unique to this design scheme.

3-55

NEW FACILITIES UNIQUE ELEMENTS
Several elements are unique to this design scheme.

3-56

EUCALYPTUS LEAVES
The dominance of the campus eucalyptus has deep roots in the College's culture.

3-57

REESTABLISHED EUCALYPTUS PATH ALONG KAPIOLANI ROAD
The new path takes design cues from its predecessor. Careful selection of a replacement species and attention to the planting details will assure a long and viable future for this design element.

3-58

ENTRY COURT FOR F. W. OLIN LIBRARY

As viewed from Leona Creek, an expanded entry court for the library will provide space for gathering and an overlook to a restored segment of Beulah Creek.

3-59

TEMPLE TO MUSIC
In deference to Howard Gilkey's original design element—the Temple to the Radio—the proposed design scheme includes a temple to music adjacent to the pond for programmed and impromptu gatherings of musicians and artists.

F. W. OLIN LIBRARY–LISSER HALL COURTYARD

The courtyard between the F. W. Olin Library and Lisser Hall and its interface with Kapiolani and Richards roads function as a significant internal campus gateway. The courtyard is a node in pedestrian and vehicular access; the campus shuttle stops in front of the courtyard on Kapiolani Road, and drivers entering the campus turn at this point to continue onto the campus. The courtyard straddles the boundary between the historic core and west campus and should exhibit the characteristics of both areas.

3-60

F. W. OLIN LIBRARY–LISSER HALL COURTYARD

Rough-hewn granite blocks of varying lengths create an informal seating area in the courtyard. Native vegetation flows from Leona Creek to Kapiolani Road.

3-61

**F. W. OLIN LIBRARY–
LISSER HALL COURTYARD**

The courtyard design blends the old with the new, the natural with the built, and the functional with the aesthetic.

3-62

**DESIGN ELEMENTS
F. W. OLIN LIBRARY–
LISSER HALL COURTYARD**

Specific design elements include new paving that incorporates the College's seal, the reintroduction of Beulah Creek near the entry to the F. W. Olin library and its expanded outdoor seating area, an accessible entrance to Lisser Hall, and lighted monuments and a bus shelter that face the intersection of Kapiolani and Richard roads.

The design accomplishes several tasks:

1. It extends the native planting of Leona Creek through the courtyard to the east side of Kapiolani Road.
2. In deference to both the formal and natural qualities of the setting, the asymmetrical shapes of California buckeye mark each of the four corners of the plaza.
3. The scheme uncovers the drainage to Leona Creek, bringing the creek closer to the campus community.
4. Pedestrian-scaled columns, inspired by the entry columns of the College, flank the courtyard's opening at Kapiolani and Richards roads.
5. A shuttle-bus shelter, with a roof inspired by the shape of eucalyptus leaves, marks the entry.
6. Pedestrian crossings raised to be flush with the adjacent sidewalk announce these areas to approaching vehicles.
7. The design uses the simplicity of scored concrete as found in the Lisser Hall portico.
8. The central plaza uses modular precast or stone paving.
9. Rough-hewn granite blocks of varying lengths create an informal seating area.
10. The College seal, surrounded by modular paving, sits in the center of the plaza.
11. An expanded entry to the F. W. Olin Library creates a generous, sun-lit outdoor area that can be furnished with tables and chairs. A light, modern balustrade helps to link the entry visually to the surrounding environment.
12. The design suggests replacing the current bridge with a bridge that is light and transparent, complementing the fine texture of plant materials and subtle sounds of Leona Creek.
13. The design provides access for the handicapped to the Lisser Hall lobby, connecting to an outdoor deck along the east side of the building.

4.
APPRECIATING THE LEGACY
HOW THE COMMUNITY BECAME AWARE OF THE COLLEGE'S LANDSCAPE HERITAGE

Mills College

Landscape Heritage Plan Lecture Series

Please join the Mills College community for a series of lectures highlighting the Mills College Landscape Heritage Plan initiative funded by the ***Getty Foundation Campus Heritage Grant Program***. Campus Heritage Grants assist colleges and universities in the United States to manage and preserve the integrity of their significant historic buildings, sites, and landscapes. The College is developing a plan that will document the campus from its 19th century origins to the present, including Mills' cultural landscape, nationally prominent architecture, and continuing commitment to innovative women's education. The following lectures are free and open to the public.

Please RSVP to Carrie Milligan
at (510) 430-2125 or cmilliga@mills.edu

Mills College is located at 5000 MacArthur Boulevard, Oakland, California 94613.

Lecture Series Calendar

The Design of Mills College: A Campus Rooted in Hope

Helen Lefkowitz Horowitz | Wednesday, October 11, 2006
Reception • 5:00 p.m.
Lecture • 5:30 p.m. with Q & A to follow
Bender Room, Carnegie Hall, Mills College

Professor and renowned author Helen Lefkowitz Horowitz of Smith College will consider Mills within the evolving plan of women's colleges in the United States and its bearing on the campus today.

Mills College and Oakland's Golden Age of Horticulture, 1918-1950

Phoebe Cutler | Wednesday, February 28, 2007
Reception • 5:00 p.m.
Lecture • 5:30 p.m. with Q & A to follow
Bender Room, Carnegie Hall, Mills College

Author Phoebe Cutler, who writes and lectures on garden and landscape history will speak about the work of landscape architect, Howard Gilkey, the first landscape architect at Mills College, and Dr. Howard McMinn noted botanist, and their influence both regionally and nationally.

Reading the Cultural Landscape of Mills College: Findings from the Getty Campus Heritage Grant Project

Vonn Marie May | Wednesday, March 28, 2007
Reception • 5:00 p.m.
Lecture • 5:30 p.m. with Q & A to follow
Bender Room, Carnegie Hall, Mills College

Vonn Marie May, Cultural Landscape Specialist and prime consultant to the Mills College Landscape Heritage Plan, will present the team's findings on the campus' development by significant architects, landscape architects, botanists, and campus planners whose influence helped shape the cultural landscape of the Mills College campus.

As part of the Getty project, Mills College developed course curricula and presented a lecture series to inform the College and public of how the campus' cultural landscape heritage significantly contributes to the College's identity.

To graduate a woman, a whole woman, is, I should say, the distinctive effort of the American college for women.

—**AURELIA HENRY REINHARDT**
TOWN & GOWN CLUB
BERKELEY, CA, 1919

COURSES

In order to incorporate the project's findings and process into the College's curriculum, two members of the Mills faculty—Professor of History Bertram Gordon, and Assistant Professor of Biology Susan Spiller—joined the team. Each designed a course offered in the spring 2007 semester. In addition, the College launched the Landscape Heritage Plan lecture series featuring four prominent historians, who presented talks on the development of Mills to members of the College community and to the public.

Mills College: Architecture and Heritage, taught by Professor of History Bertram Gordon, focused on Mills College from its inception as a seminary in 1852, through the establishment of the College in 1885, to the present. Professor Gordon placed the development of Mills in a larger historical context of culture, education, and architecture in the modern Western world.

A key component of the course was the development of student research and archival skills through the use of historical documents and photographs related to Mills and its history. Borrowing the approach of the French historian Pierre Nora, Professor Gordon asked students to write papers on a "site of memory" (*lieu de mémoire*), a place with significance in the history of Mills. They could select sites anywhere on the Mills campus or even off campus if they related to the history of the College. Projects focused not only on the physical history of a site but, more importantly, its historical significance to the College community–namely, what kinds of memories and images were associated with the site. Students visited the Mills College archives in the library's Special Collections department early in the semester and used the collections, together with oral and email interviews they conducted,

4-1

LECTURE INVITATION
Mills College distributed invitations to alumnae, the College community, and historic and design professionals throughout the San Francisco Bay Area.

to expand their examination of various sites of interest on campus. Leah Albin, a teaching assistant for the class who was also employed in the Mills library archive, was able to help students with specific projects and provide technical training in archival work. Students tended to select sites that were closely related to their own lives on campus, such as dormitories, or sites that were readily identifiable as significant landmarks on campus.

Fifteen undergraduate students enrolled in the class and wrote about a wide variety of topics, including campus structures such as the Ethel Moore and Mary Morse residence halls, El Campanil, and other facilities with historic significance in the development of the institution. Some students researched campus landscape topics, including eucalyptus trees, Lake Aliso, and Sunnyside, where the College's founders are buried. Other papers explored broader historical topics in relation to the evolution of the Mills campus, including the removal of the Julia Morgan gymnasium, demolished in the middle of the 20th century; religion at Mills, addressing the succession of buildings used as a campus chapel; the Art Gallery; and the offices of the student newspaper, the *Mills College Weekly*, where five students in the class worked.

Field trips and guest speakers supplemented class lectures. One of these was a visit to the Special Collections Library, where Mills Archivist Janice Braun talked about research methods and provided the students with an overview of documents and photographs in the collection. The class' outdoor campus field trips included two historic walking tours, a visit to the top of El Campanil, and a tour of the newly constructed Natural Sciences Building. The class made an off-campus visit to the Oakland Museum to study California history as exhibited there. At the museum the class focused on the early history of California as well as on how a museum stages its exhibits and fulfills its educational mission.

Guest speakers for the class included Janet Holmgren, President of the College, who spoke about Mills history and how it relates to her vision for the College's future;

Marianne Sheldon, Associate Provost and Professor of History, who addressed gender roles and feminism in late 20th-century America; Jane King, an alumna from the class of 1942, who talked about student life on campus during the late Depression and early war years; and Judy Silva, Director of Marketing, who talked about the College's image and reputation in the larger world. Karen Fiene, the campus architect, along with Vonn Marie May and Robert Sabbatini, the College's Getty Grant consultants, also visited the class to discuss buildings; landscape; and past, present, and future campus planning with students. Paul Richards, former Director of Campus Facilities, met with the class as well to discuss issues related to the College's landscape and restoration of Lake Aliso. In addition to completing their research projects, students took a midterm and a final exam, each composed of short-answer and essay questions.

The Greening of Mills College: How to Conserve Resources for a Sustainable Future was an interdisciplinary sophomore core course taught by Assistant Professor of Biology Susan Spiller, focusing on the environment and sustainability on the

4-2

HOWARD MCMINN WITH STUDENTS, CIRCA 1937
Mills College's commitment to using its campus as a teaching laboratory dates back many decades. Professor of Botany Howard McMinn's intensive study and education about California native plant material led him to write several authoritative publications.

Mills campus and its broader urban watershed. The course involved seminar-style discussion, lectures, guest speakers, and field trips across the Mills campus, in local neighborhoods, and at the Leona Creek watershed on campus. Audiovisual tools were used to tape events and interviews. Interdisciplinary topics related to environment, ecology, culture, and policy.

The class began with a van trip to a nearby restored watershed, Arroyo Viejo, and then to a ridge of the East Bay Hills overlooking Mills to get a bird's eye view of the Mills campus. After this, the class enjoyed presentations from a variety of Mills staff and faculty, as well as visiting speakers with expertise in sustainability practices, including the lead consultants to the Getty-funded Landscape Heritage Plan, who gave students a historical perspective of the Mills campus. The students also attended the Landscape Heritage Lecture Series during the spring semester.

Karen Fiene, campus architect; Louise Gilbert, from the College's development office; Karen Maggio, Associate Vice President for Campus Planning and Facilities; and Vonn Marie May, a cultural landscape specialist, made presentations to the class about the heritage of the College's landscape and the process involved in the College's application to the Getty Foundation. Students began to understand the importance of publishing the Landscape Heritage Plan as a vital tool that will help continue to define the importance of the campus' natural and built environments.

The 21 students in this seminar-style class benefited from lively discussions with guest speakers and among themselves, which helped define their final project topics. Professor Spiller's goal for the class was to stimulate thoughts about issues of sustainability on college campuses, with the wider hope that interesting and provocative projects might help the Mills campus move toward sustainable campus practices. The students' projects are available on the *Greening of Mills College* website at *http://courses.mills.edu/courses/public/coll060m.*

4-3

LECTURE VIDEOS, 2007
Mills College created videotape records of the well-attended lecture series.

LECTURE SERIES

The Landscape Heritage Plan lecture series brought experts to campus to discuss the impressive historical and cultural impact made by notable individuals involved in the development of the College's architecture and landscape and its cultural significance as a women's educational institution during specific historical periods. Each lecture was attended by 50 to 75 people from the Bay Area, including Mills College students and faculty, campus administrators, students and faculty from other campuses, architects and historians, and members of the public.

Summarized on the following pages, the Landscape Heritage lecture series at Mills has stimulated increased public awareness about the rich history of Mills and has garnered great interest among historic, architectural, and preservation communities.

4-4

CARNEGIE HALL DEDICATION, 1906

Julia Morgan's Margaret Carnegie Library was named and funded by the steel king, Andrew Carnegie, as a tribute to his young daughter.

4-5

CARNEGIE HALL INTERIOR, CIRCA 1916

This second-floor grand space (the Bender Room from 1954 to present) once housed 7,600 volumes and stands today as an exquisite display of muscular redwood trusses, paneling, and built-in bookcases.

The Design of Mills College: A Campus Rooted in Hope

Helen Lefkowitz Horowitz

A stark contrast emerges in comparing the early design of colleges meant for men and those meant for women. Each has its own social narrative and architectural derivations, and the differences remind us of what was once the dominant attitude of Americans about the higher education of women.

English universities were based on a monastic model of the quadrangle and cloister. Conversely, early American colleges were designed on an open plan: an "Old College" building fronted by a generous yard with other detached buildings. From the outset, the distinctive feature of the American New England college was its openness to society. Most importantly, there was no chapel on-site. Harvard's plan was confirmed by Yale, which put its buildings on the New Haven Green.

As new buildings were added to Harvard and Yale, the academical village took shape. The college was to be a community within a community, containing provisions for learning and living, with the exception of worship. It was to front an important street and have solid, impressive buildings. The buildings were to be separate, with the "yard" initially given over to a common ground.

When we turn to the first colleges for women, we find that different rules went into effect. The building type was distinctive. The model on which they were first based was not the male New England college, but the academy or seminary.

Mary Lyon, founder of Mount Holyoke in 1837, saw the seminary as offering curriculum beyond the English model. She saw it as changing the consciousness of women students. In establishing the architectural precedent of her day for such institutions, she built the seminary using design principals of a mental asylum, a system of order and isolated setting, designed to create internal order in the minds of

Helen Lefkowitz Horowitz, the Sydenham Clark Parsons Professor of History at Smith College, does research that ranges over urban life, cultural philanthropy, sexuality, biography, and women in higher education. She is author of *Alma Mater: Design and Experience in the Women's Colleges from their Nineteenth-Century Beginnings to the 1930s* (Amherst: University of Massachusetts Press, 1993). On October 11, 2006, Dr. Horowitz lectured on "The Design of Mills College: A Campus Rooted in Hope." She probed the ways in which founders of women's colleges expressed their hopes and fears about women as they offered them the liberal arts.

its inmates. Nineteenth-century reformers believed that if you separated those who were disordered in their minds and placed them in a structure of external order, they would internalize the rules and develop an inner psychic order.

Following this asylum pattern and adopting its rhetoric of family government, a seminary building ordered relationships hierarchically. With a central entrance and stairwell and provisions for living, learning, and working, Mount Holyoke held no places for retreat, no interstices for freedom.

4-6

MILLS COLLEGE, 1870s
A building presumably erected for a seminary but with grandeur clearly designed for a college.

The first true college for women in the United States was Vassar College planned in 1861. Matthew Vassar decided to create a first real college for women, "to be to them, what Harvard and Yale are to young men." The college had an undiluted liberal arts course and full faculty; but unlike its counterparts for male students, the building design followed the form of a female seminary. When Wellesley College came into being in 1875, it took much the same form as Vassar. As with Mills, Wellesley and Vassar were sited in a picturesque landscape setting. A look at the early architecture of Vassar and Wellesley shows us that Mills Hall is a hybrid of structures built at sister institutions on the east coast.

In the 1870s, Smith College seemed to be an expression of fear, in this case of assertive women of the suffrage era. In the East, concerns over college life in a huge seminary building and a culture of strong-mindedness led Smith to place students in homelike settings. This domestic approach found an even greater expression in the early Radcliffe design, which suggests that fear, rather than hope, dominated planning in the new women's colleges in the two decades after 1875.

When Bryn Mawr began to build, it took the form of quadrangles, one of the earliest and most elegant compositions of College Gothic. Here was the critical breakthrough in the building history of women's colleges. Before Bryn Mawr, women's colleges had distinctive structures designed to protect young women as they received higher learning. At Bryn Mawr, buildings at women's colleges entered into the mainstream of collegiate architecture, at least on the outside.

Mills Hall, erected in 1871, expressed Cyrus and Susan's ultimate intentions. Susan's mentor Mary Lyon, who may not have approved of its fashionable Second Empire style, would have applauded the internal design. Yet, Mills Hall serves as evidence that the founders hoped their school would become a college.

In 1916 Aurelia Henry Reinhardt, Mills College's new president, ushered in a period of dynamic change. A new professionalism quickly came into play, emphasizing the importance of Mills in training "teachers and professional workers into every field." The genius of Walter H. Ratcliff's campus designs is his placement of new buildings to respect the site, situating them to accept the campus' hills.

Considered as a whole, the structures designed by an array of prominent architects—Julia Morgan, Bakewell & Brown, and Ratcliff—residence halls, classroom buildings, facilities for student life, even a lost gymnasium create a gracious campus in the California idiom. In examining the Mills College campus that was built in the late 19th and early 20th centuries, important aspects are present:

- Mills Hall: A building presumably erected for a seminary but with grandeur clearly designed for a college.
- Attention to a ceremonial structure, such as Julia Morgan's El Campanil. It expresses the seriousness, dignity, and strength of Mills College to the world outside.
- In its great building phase, the construction of a magnificent composition of Mediterranean buildings sited with an understanding of Mills's unique landscape of small hills and extraordinary mature vegetation.

All of these suggest that in the planning of Mills through the Second World War, it is hope rather than fear that shaped the campus. Mills shows a more consistent pattern of hope, promotion, and expansiveness than do earlier women's colleges. This optimism suggests a college that embraced the strengths and talents of women. As Mills enters a new phase with the Mills College Landscape Heritage Plan, the College comes to a new epoch bringing together the pieces of the past to create a renewed whole.

Mills Horticultural Glory Years: 1915–1940

Phoebe Cutler

Howard McMinn (1891–1963), Botanist and Author

Howard Gilkey (1890–1972), Campus Landscape Architect

Two plantsmen, the botanist Howard McMinn and the landscape architect Howard Gilkey, contributed to the physical and aesthetic nature of the campus while establishing the College as a force for spreading the knowledge and appreciation of both native and exotic vegetation. During a time that is now regarded as the golden age of horticulture in the Bay Area, both men were notables in their fields based on their work at Mills College.

McMinn and Gilkey were originally from the Midwest, born within a year of each other. Both graduated from the University of California, Berkeley; McMinn, with a master's degree.

McMinn's tenure at Mills College lasted for more than four decades. He was in residence from 1916 through the early 1960s. As professor of botany, his commitment to the study and education of California native plant material led him to write several authoritative publications. His scholarly work was directly related to his botanical experiments conducted in outdoor classrooms, as he and his students planted more than 275 different native species. Offsetting the rampant growth of exotic tree canopy, McMinn also planted hundreds of redwood, spruce, and Douglas fir. His influence can still be felt campuswide. McMinn began the transition from the late Victorian exotic horticultural boom through a mid 20th-century sensibility and appreciation for California native plants.

Most of Gilkey's work was executed predominantly in the 1920s as the campus landscape architect, although his tenure lasted for nearly twenty years off and on from 1915 through the 1930s. He worked with the campus architect Walter

Phoebe Cutler, a landscape historian, lectures and writes on garden and landscape history. The focus of her work has been the U.S., 1890 to 1950, as represented by *The Public Landscape of the New Deal* (New Haven: Yale, 1986) and her recent essay, "California and the Municipal Rose Garden Movement," *Eden*, Spring 2005. Her lecture on February 28, 2007, "Mills Horticultural Glory Years: 1915–1940," addressed the work of the campus landscape architect Howard Gilkey and of Dr. Howard McMinn, who chaired the College's botany department from 1946 to 1956 and who planted 2,500 native trees representing 41 species at Mills College. McMinn documented his experimental horticultural work at Mills, producing scholarly reference publications used throughout California and the West.

H. Ratcliff, Jr., on project-specific designs that complemented Ratcliff's "modern" European revival architecture.

His early training at Berkeley's Division of Landscape Gardening and Floriculture exposed him to a wide range of 17th- and 18th-century classical gardens of Europe. Gilkey's early years were spent working with the horticulture legends Luther Burbank and Carl Purdy. For two years he assisted the bulb and wildflower specialist Purdy with landscape design and set-up of the exhibitions of the Palace of Horticulture and its South Gardens at the Pan-Pacific Exposition of 1915 in San Francisco.

The same year Gilkey was working on the fair, he assisted the architects Charles Dickey and John Donovan in preparing the first master plan for Mills College. Although the plan was never implemented, it became Gilkey's entrée onto the campus and led to full-time landscape design projects.

4-7

ETHEL MOORE FOUNTAIN, 1950s
Designed by Howard Gilkey, campus landscape architect.

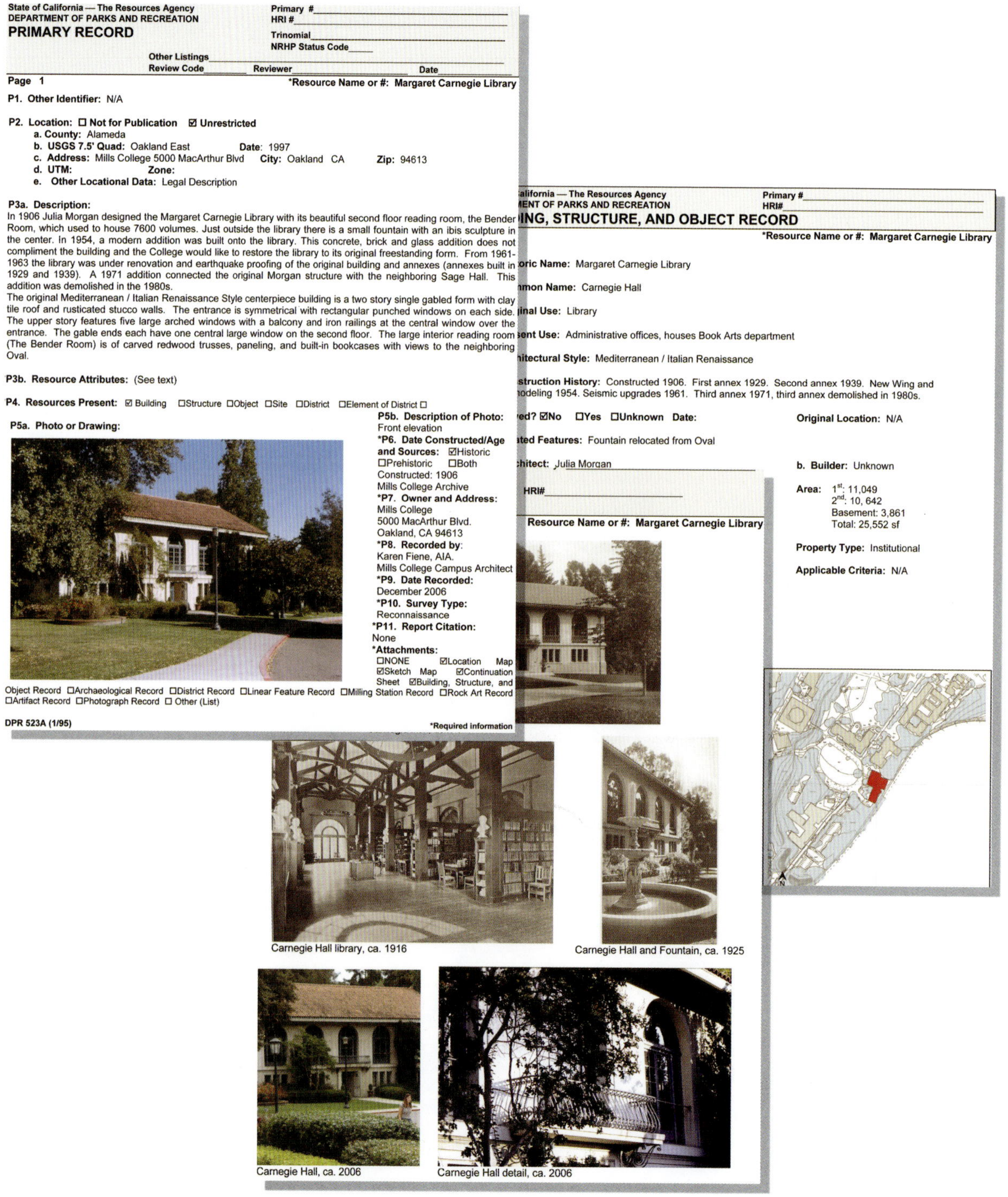

State of California — The Resources Agency
DEPARTMENT OF PARKS AND RECREATION
PRIMARY RECORD

Primary #
HRI #
Trinomial
NRHP Status Code
Other Listings
Review Code Reviewer Date

Page 1 *Resource Name or #: Margaret Carnegie Library

P1. Other Identifier: N/A

P2. Location: ☐ Not for Publication ☑ Unrestricted
a. County: Alameda
b. USGS 7.5' Quad: Oakland East Date: 1997
c. Address: Mills College 5000 MacArthur Blvd City: Oakland CA Zip: 94613
d. UTM: Zone:
e. Other Locational Data: Legal Description

P3a. Description:
In 1906 Julia Morgan designed the Margaret Carnegie Library with its beautiful second floor reading room, the Bender Room, which used to house 7600 volumes. Just outside the library there is a small fountain with an ibis sculpture in the center. In 1954, a modern addition was built onto the library. This concrete, brick and glass addition does not compliment the building and the College would like to restore the library to its original freestanding form. From 1961-1963 the library was under renovation and earthquake proofing of the original building and annexes (annexes built in 1929 and 1939). A 1971 addition connected the original Morgan structure with the neighboring Sage Hall. This addition was demolished in the 1980s.
The original Mediterranean / Italian Renaissance Style centerpiece building is a two story single gabled form with clay tile roof and rusticated stucco walls. The entrance is symmetrical with rectangular punched windows on each side. The upper story features five large arched windows with a balcony and iron railings at the central window over the entrance. The gable ends each have one central large window on the second floor. The large interior reading room (The Bender Room) is of carved redwood trusses, paneling, and built-in bookcases with views to the neighboring Oval.

P3b. Resource Attributes: (See text)

P4. Resources Present: ☑ Building ☐Structure ☐Object ☐Site ☐District ☐Element of District ☐

P5a. Photo or Drawing:

P5b. Description of Photo: Front elevation
***P6. Date Constructed/Age and Sources:** ☑Historic ☐Prehistoric ☐Both
Constructed: 1906
Mills College Archive
***P7. Owner and Address:** Mills College
5000 MacArthur Blvd.
Oakland, CA 94613
***P8. Recorded by:** Karen Fiene, AIA.
Mills College Campus Architect
***P9. Date Recorded:** December 2006
***P10. Survey Type:** Reconnaissance
***P11. Report Citation:** None
***Attachments:** ☐NONE ☑Location Map ☑Sketch Map ☑Continuation Sheet ☑Building, Structure, and Object Record ☐Archaeological Record ☐District Record ☐Linear Feature Record ☐Milling Station Record ☐Rock Art Record ☐Artifact Record ☐Photograph Record ☐ Other (List)

DPR 523A (1/95) *Required information

...alifornia — The Resources Agency
...MENT OF PARKS AND RECREATION
...ING, STRUCTURE, AND OBJECT RECORD
Primary #
HRI#
*Resource Name or #: Margaret Carnegie Library

...oric Name: Margaret Carnegie Library
...mon Name: Carnegie Hall
...inal Use: Library
...ent Use: Administrative offices, houses Book Arts department
...itectural Style: Mediterranean / Italian Renaissance
...struction History: Constructed 1906. First annex 1929. Second annex 1939. New Wing and ...odeling 1954. Seismic upgrades 1961. Third annex 1971, third annex demolished in 1980s.
...ed? ☑No ☐Yes ☐Unknown Date: Original Location: N/A
...ted Features: Fountain relocated from Oval
...hitect: Julia Morgan b. Builder: Unknown
Area: 1st: 11,049
2nd: 10, 642
Basement: 3,861
Total: 25,552 sf
Property Type: Institutional
Applicable Criteria: N/A

HRI#
Resource Name or #: Margaret Carnegie Library

Carnegie Hall library, ca. 1916
Carnegie Hall and Fountain, ca. 1925
Carnegie Hall, ca. 2006
Carnegie Hall detail, ca. 2006

4-8

HISTORIC BUILDING DOCUMENTATION

As part of this study, Mills College staff documented 27 buildings and built landscape features.

Reading the Cultural Landscape of Mills College: Historic Findings from the Getty Foundation Campus Heritage Project

Vonn Marie May

The major goal of reading the landscape of Mills College is to produce a document that identifies the historic, character-defining features of the campus to better inform future campus planning. The landscape heritage plan for Mills College must reconcile more than a century's worth of landscape and architectural growth. It must also provide cues from the founders, notable designers and planners, design movements, and introduction of exotic plant materials.

The team produced a chronological narrative of the built environment from 1868 to 1949 that documents the history of the physical development of the campus. Two periods of historic significance were identified: 1868 to 1916, the period of Dr. Cyrus and Susan Mills, the founders, which was the English country estate/picturesque era; and 1916 to 1949, the period of Dr. Aurelia Henry Reinhardt, President of Mills, with Walter H. Ratcliff, Jr., the campus architect, and the rise of the Spanish colonial revival/modern era.

Surveys and physical assessments of existing historic features were conducted, which included the evaluation of spatial organization, land use patterns, context, historical architecture, topography, historic tree palette, circulation patterns, hardscape features, natural systems, views and vistas, and overall site integrity.

Two high-profile design initiative sites were explored for the application and synthesis of the historic findings. Through this process, the College drafted two designs that will benefit the preservation and interpretation of Mills College in future projects.

The methods used in this effort were developed and publicized by the National Park Service through its extensive technical documentation for cultural resource

Vonn Marie May is a highly regarded cultural landscape specialist and a prime consultant to the Mills College Landscape Heritage Plan. She also consulted on the *Landscape Heritage Plan University of California, Berkeley*, funded by the Getty Foundation. Her expertise focuses on historical resources and cultural landscapes. In her lecture on March 28, 2007, titled "Reading the Cultural Landscape of Mills College: Historic Findings from the Getty Foundation Campus Heritage Project," she presented the findings from her survey and assessment work on the campus and a chronological narrative of campus history. Her assessment of the campus' cultural landscape identified two periods of historic significance and character-defining features. She guided College staff in their preparation of campus-wide surveys, which documented significant architects, landscape architects, botanists, and campus planners, whose influence helped to shape the cultural landscape of the campus during the most significant periods in the College's history.

4-9

SAGE HALL, 1910s
Sarah Sage Hall Library and art gallery with parterre garden.

4-10

PATIO OF THE PINES, 1925
Courtyard design for Ethel Moore Hall by Howard Gilkey, landscape architect. Drawn by Vernon Morse.

investigations. In particular, the project team used the Secretary of the Interior's *Standards for the Treatment of Historic Properties with Guidelines for the Treatment of Cultural Landscapes*, and *National Register Bulletin #15: How to Apply the National Register Criteria for Evaluation*. The National Register of Historic Places evaluates properties 50 years and older based on specific criteria.

Overall observations from the study indicate that the Mills College campus landscape has gone beyond its original design intent. Yet, despite the predominance of exotic species and the loss of significant architecture and open spaces, it continues to convey historic integrity from both its periods of significance. An understanding of the existing campus landscape, including elements that define a cultural landscape and cues from the College's origins, should be a primary factor to consider in future planning.

4-11

KAPIOLANI ROAD, 1947

Pedestrians and bicyclists moved freely along Kapiolani Road prior to the heavier automobile traffic experienced today.

4-12

MUSIC BUILDING, 1931
Walter H. Ratcliff, Jr., architect.

4-13

ART GALLERY
The primary entry for the Art Gallery was on the north near the old K Leona station. Designed by Walter Ratcliff, Jr., in 1925.

4-14

DESIGN STUDY FOR CHAPEL
Design study of the Mills College Chapel prepared by Walter H. Ratcliff, Jr., in 1933.

Architect Walter H. Ratcliff, Jr., at Mills College

Woodruff Minor

Excerpted from *The Architecture of Ratcliff*, by Woodruff Minor

Walter H. Ratcliff, Jr. (1881–1973), came to California from England in 1894. His father, a clergyman and teacher, moved the family to Berkeley several years later so that his three children could attend the University of California. The town was fast becoming the center of architectural achievement, exposing the young man to an array of work that set him on the path to his career. Following his graduation from the University of California, Berkeley, in 1903, and with no formal training in architecture, young Ratcliff worked and apprenticed in the office of the University's campus architect, John Galen Howard. Ratcliff then studied for a time in Europe and entered a short-lived partnership in San Francisco before establishing the architectural practice in Berkeley in 1906 that would be the focus of his 50-year career. He soon took his place among Berkeley's well-known coterie of architects, including Bernard Maybeck and Julia Morgan, initially as a designer of arts-and-crafts houses and ultimately as an exponent of revivalism, deploying a variety of styles on a wide range of commercial, civic, academic, and ecclesiastical commissions.

Walter Ratcliff's appointment as campus architect of Mills College in 1923 came at an opportune time. The 40-year-old architect had four young children to support and had recently been dismissed as city architect of Berkeley. The new position would provide a steady flow of projects over the coming decade during a sluggish post-World War I economy. In all, Ratcliff's association with Mills College would last 23 years, one year longer than John Galen Howard's tenure as supervising architect at the University of California.

Since its opening two years before the University of California, the Mills campus consisted of its centerpiece, Mills Hall, a sprawling wood structure erected in 1871 in the Second Empire mansard style, and a scattering of other Victorian buildings,

Woodruff Minor is an architectural historian who has written extensively on Bay Area history. His most recent book is *The Architecture of Ratcliff* (Berkeley, CA: Heyday Books, 2006). He is a graduate in history from the University of California, Berkeley, and holds a master's degree in urban planning from the University of Oregon. He has been working in the Bay Area since 1977, initially as a preservation planner and later serving as Chair of the Landmarks Board and Assistant Planner for the City of Alameda. He has more than 35 years of consulting experience as well, working on such projects as the California Environmental Air Quality Act and assessment for the National and California Registers of Historic Buildings. He currently serves as consultant on projects for the Port of Oakland Army Base reuse and the San Francisco Public Utilities Commission for the Hetch Hetchy Reservoir and Aqueduct. His lecture on April 19, 2007, concentrated on the work of the architect, Walter H. Ratcliff, Jr., who served as Mills Campus Architect from 1923 to 1946.

which dotted the grounds. Around Mills Hall was a distinguished group of structures designed by Julia Morgan. Following the death of Susan Mills, the College's co-founder and longtime president, commissions began passing to a variety of notable firms. By World War I, the campus presented a mix of Victorian and Mission/Mediterranean eclecticism.

With the appointment of its new president, Aurelia Henry Reinhardt, in 1916, Mills embarked on a program of expansion to accommodate growing enrollment. In 1915, Oakland architects Charles W. Dickey and John J. Donovan had produced the College's first master plan. This plan, however, did not appeal to the most powerful Mills Trustee, Phoebe Hearst, who privately commissioned an alternate scheme from Bernard Maybeck. His scheme, with elegant layouts of boulevards, plazas, and monumental buildings recalling a beaux-arts spirit, envisioned an entirely new campus, treating the site like a blank canvas with no regard for existing structures except for Morgan's bell tower. Though the Board formally adopted Maybeck's plan in 1919, it was never realized, due in part to the death of Mrs. Hearst later that year.

4-15

RIDING STABLES,1935
Now demolished, the riding stables were located in the northwest corner of the campus adjacent to Seminary Avenue; Walter H. Ratcliff, Jr., architect.

In 1921, the Trustees turned to Walter Ratcliff for yet another iteration of the campus. Submitted that same year, his master plan for Mills College reprised Maybeck's beaux-arts concepts in its salient features, proposing an idealized axial ensemble for the site. While important pieces of Ratcliff's plan were brought to completion, it remained largely unrealized. The unsurprising decision by the Board of Trustees to leave existing facilities in place, in particular Mills Hall and the surrounding Julia Morgan structures, meant that the contemplated main axis and plaza would never be completed.

Aurelia Reinhardt had persuaded the Trustees that new campus buildings should adhere to the Spanish Colonial Revival style, and Ratcliff's work for the College followed this line, with touches of the beaux-arts classicism reminiscent of the Berkeley campus. During his tenure at Mills, Ratcliff and his firm completed a variety of campus projects, including:

- the Art Gallery (1925)
- the Music Building (1928), which garnered widespread notice and received the highest award at the 1929 biennial exhibition of the Northern California Chapter of the American Institute of Architects
- a dormitory complex (1926–1944), built in three phases, which housed over 200 dormitory rooms and a half-dozen living and dining rooms
- Wetmore Gate (1925), which served as the main entrance to campus off of Seminary Avenue, adjoined by a lodge and circular fountain
- Richards Gate (1933) on the west side of campus
- the Norman Bridge Health Center (1930)
- Faculty Village (1930s–1940s), a group of nine single-family duplex houses

After the retirement of Aurelia Henry Reinhardt, the new College administration turned to a more modernist approach to campus design, and in 1946 Ratcliff's tenure as supervising architect ended.

SOURCES

Bancroft, Hubert Howe. *The Works, Vol. XXIV: History of California 1860–1890.* San Francisco: History Company, 1890.

Bagwell, Beth, *Oakland: The Story of a City.* Oakland Heritage Alliance, 1994

Beatty, Russell *A. Vegetation Management Plan for Mills College.* Oakland: Mills College, 1996.

Boutelle, Sara Holmes. *Julia Morgan: Architect.* New York: Abbeville Press, 1995.

Downing, A. J. *A Treatise on the Theory and Practice of Landscape Gardening, Adapted to North America.* New York: Wiley & Putnam, 1841.

Garrison, Myrtle. *Romance and History of California Ranchos.* San Francisco: Harr Wagner Publishing Company, 1935.

Hedley, George. *Aurelia Henry Reinhardt: Portrait of a Whole Woman.* Oakland: Mills College, 1961.

Herrada, Mario. Mills College: *The Landscape Preservation of Richards Road.* Master's thesis, University of California, Berkeley, 1996.

Hinkel, Edgar, J., and William E. McCann, eds. *Oakland 1852–1938: Some Phases of the Social, Political, and Economic History of Oakland, California,* Vol. I. Oakland: Oakland Public Library, 1939.

Horowitz, Helen Lefkowitz. *Alma Mater: Design and Experience in the Women's Colleges from Their 19th Century Beginnings to the 1930s.* New York: Alfred A. Knopf, 1984.

Jackson, John Brinckerhoff. *American Space: The Centennial Years 1865–1876.* New York: W. W. Norton, 1972.

James, Elias Olan. *The Story of Cyrus and Susan Mills.* Stanford: Stanford University Press, 1953.

Jellicoe, Geoffrey and Susan Jellicoe. *The Landscape of Man: Shaping the Environment from Prehistory to Present Day.* London: Thames and Hudson, 1987.

Kasapligil, Baki. *Trees and Shrubs of Mills College.* Oakland: Mills College, 1967.

Keep, Rosalind A. *Fourscore and Ten Years: A History of Mills College.* Oakland: Mills College, 1946.

LTD Engineering Inc. and LSA Associates Inc. *Lion Creek Stream Channel Plan, Mills College California, 2007.*

McAlester, Virginia and Lee. *A Field Guide to American Houses.* New York: Alfred A. Knopf, 2000.

Minor, Woodruff. *The Architecture of Ratcliff.* Berkeley: Heyday Books, 2006.

Newton, Norman T. *Design on the Land: The Development of Landscape Architecture.* Cambridge, MA: Belknap Press of Harvard University Press, 1973.

Reiss, Suzanne. *Julia Morgan Architectural History Project Interviews: Oral History Transcript and Related Material,* 1974–1976, Vol. II. Berkeley: Regional Oral History Office, The Bancroft Library, University of California, Berkeley, 1976.

Sanford, T. E. *The Architecture of the Southwest.* Tucson: University of Arizona Press, 1997.

Starr, Kevin. *Americans and the California Dream: 1850–1915.* New York: Oxford University Press, 1973.

Streatfield, David C. *California Gardens: Creating a New Eden.* New York: Abbeville Press, 1994.

Weitze, Karen J. *California's Mission Revival.* Los Angeles: Hennessey & Ingalls, 1984.

Periodicals

Various dates and authors:

Oakland Tribune

Pacific Coast Architect

The Architect and Engineer

Mills College Weekly

Mills Quarterly

Technical Periodicals

Birnbaum, Charles A. *Preservation Brief 36: Protecting Cultural Landscapes: Planning, Treatment and Management of Historic Landscapes.* Washington, D.C.: U.S. Department of the Interior, National Park Service, 1994.

Birnbaum, Charles A., ed. With the assistance of Christine Capella Peters. *The Secretary of the Interior's Standards for the Treatment of Historic Properties with Guidelines for the Treatment of Cultural Landscapes.* Washington, D.C.: U.S. Department of the Interior, National Park Service, 1996.

Keller, J. Timothy and Genevieve P. Keller. *National Register Bulletin: How to Evaluate and Nominate Designed Historic Landscapes.* Washington, D.C.: U.S. Department of the Interior, National Park Service, 1987.

National Park Service. *National Register Bulletin 15: How to Apply the National Register Criteria for Evaluation.* Washington, D.C.: U.S. Department of the Interior, National Park Service, 1995.

Page, Robert R., Cathy Gilbert, and Susan Dolan. *A Guide to Cultural Landscape Reports: Contents, Process, and Techniques.* Washington, D.C.: U.S. Department of the Interior, National Park Service, 1998.

Repositories

California Polytechnic State University, The Julia Morgan Collection, Special Collections Department, Robert E. Kennedy Library, San Luis Obispo

F. W. Olin Library, Mills College, Oakland, California

Map Files, Facilities Department, Mills College, Oakland, California

College of Environmental Design, University California, Berkeley, Berkeley, California

The Bancroft Library, University of California, Berkeley, Berkeley, California

History Room, Oakland Public Library, Oakland, California

Illustration Credits and Permissions

Back cover, Art Zendarski

0-1, Eadweard Muybridge (published by Bradley & Rulofson)

1-6, Milton Mann Studios

1-7, Robert Sabbatini

2-1, Eadweard Muybridge (Helios)

2-2–2-3, Robert Sabbatini

2-4, Rieman & Co.

2-5, Eadweard Muybridge

2-6, Isaiah W. Taber

2-7, Edwin Graybiel

2-8, Isaiah W. Taber

2-13, Ted Streshinsky/CORBIS

2-15, Isaiah W. Taber

2-18, Roi Partridge

2-21, Lothers & Young

2-22, Reproduced by permission from the Julia Morgan Collection, Special Collections, California Polytechnic State University.

2-23, R. J. Waters.

2-24, Roi Partridge

2-25, J. Stewart

2-28, Roi Partridge
2-29, Commercial & Photo View Co.
2-35, Ted Streshinsky
2-36, Robert Sabbatini
2-37, Young's Photography
2-38, Gabriel Moulin. Copyright Moulin Studio Archives
2-41, Reproduced by permission from the Ratcliff Archive
2-43, Gabriel Moulin. Copyright Moulin Studio Archives
2-44, Photograph by Imogen Cunningham. Copyright, the Imogen Cunningham Trust
2-47, Photograph by Imogen Cunningham. Copyright, the Imogen Cunningham Trust
2-48, George Derbfus
2-50, *Mills Quarterly*, October 1926, v. IX, n. 3
2-51, Don Jones
2-52, Photograph by Imogen Cunningham. Copyright, the Imogen Cunningham Trust
2-58, John Wright
2-61, George Tagney
2-62, Robert Sabbatini
3-1, Art Zendarski
3-2–3-4, Robert Sabbatini
3-5, Vonn Marie May
3-6–3-10, Robert Sabbatini
3-11–3-12, Roi Partridge
3-13, Robert Sabbatini
3-15, Robert Sabbatini
3-19–3-20, Robert Sabbatini
3-21, Vonn Marie May
3-23–3-25, Robert Sabbatini
3-26–3-30, Art Zendarski
3-31–3-35, Robert Sabbatini
3-36, Karen Fiene
3-37–3-40, Robert Sabbatini
3-41, Vonn Marie May
3-42, George Tagney (Allendale Photo)
3-43–3-49, Robert Sabbatini
3-50–3-52, Art Zendarski
3-53–3-55, Robert Sabbatini
3-56–3-61, Art Zendarski
3-62, Robert Sabbatini
4-1, Carrie Milligan
4-3, Robert Sabbatini
4-6, Eadweard Muybridge
4-8, Brian Harrington
4-10, Mills Quarterly, October 1926, v. IX, n. 3
4-11, Photograph by Imogen Cunningham. Copyright, the Imogen Cunningham Trust
4-12, Gabriel Moulin. Copyright Moulin Studio Archives
4-14, Reproduced by permission from the Ratcliff Archive.

All efforts were made to identify, locate, and acknowledge the copyright holders for the photographs that appear in this work that are not in the public domain.

ACKNOWLEDGEMENTS

We are proud to dedicate *Celebrating the Cultural Landscape Heritage of Mills College* to former Vice President for Finance and Administration and Treasurer Elizabeth Burwell, who retired in May 2007 after 38 years at Mills College, for her unwavering vision, inspiration, and support for this plan.

Mills College wishes to thank the Getty Foundation for its generous Campus Heritage Grant and especially Robert Z. Melnick, Visiting Senior Program Officer, for his thoughtful guidance.

Mills College Project Team

Leah Albin, Mills College Student Assistant
Janice Braun, Special Collections Curator and Director, Center for the Book, F. W. Olin Library
Elizabeth Burwell, former Vice President for Finance and Administration and Treasurer
Dwayne De Fount, former Grounds Manager, Campus Facilities
Karen Fiene, AIA, Campus Architect
Louise Gilbert, Director of Development for Foundations and Corporations, Office of Institutional Advancement
Bertram Gordon, Professor of History
Brian Harrington, Architecture Assistant
Renee Jadushlever, Vice President for Operations
Holly Kernan, News Director, KALW Public Radio and Mills College Visiting Scholar and Lecturer
Karen Maggio, Associate Vice President for Campus Planning and Facilities
Stephanie L. Mazow, Assistant Director for Foundation and Corporate Relations, Office of Institutional Advancement
Carrie Milligan, Executive Assistant, Office of the Vice President for Finance and Administration and Treasurer
Karma Pippin, Assistant Archivist, F. W. Olin Library
Paul Richards, former Director of Facilities
Virginia V. Rivera, CFRE, CSPG, Vice President for Development
Marianne Sheldon, Associate Provost and Professor of History
Judy Silva, Assistant Vice President for Marketing
Susan Spiller, Assistant Professor of Biology
Andrew Workman, Associate Provost for Academic Affairs

Principal Project Consultants

Vonn Marie May, Cultural Landscape Specialist
Robert Sabbatini, AICP ASLA
Art Zendarski, Illustrator, Zendarski Studio

Participating Consultants

Jim Clark, PhD, Consulting Arborist, HortScience
Peter Ehrlich, Forestry Manager, The Presidio Trust
Ron Lutsko, Jr., Principal, Lutsko Associates Landscape
Betty Marvin, Historic Preservation Planner, City of Oakland

Landscape Heritage Plan Lecture Series

Phoebe Cutler, Landscape Historian
Helen Lefkowitz Horowitz, The Sydenham Clark Parsons Professor of History, Smith College
Vonn Marie May, Cultural Landscape Specialist
Woodruff Minor, Architectural Historian

INDEX

Numbers in italics refer to a caption and/or illustration.